Release 8

Introductory Guide
Second Edition

*"The real voyage of discovery consists not in seeking new
landscapes, but in having new eyes."*

Marcel Proust

JMP, A Business Unit of SAS
SAS Campus Drive
Cary, NC 27513

8.0.2

The correct bibliographic citation for this manual is as follows: SAS Institute Inc. 2009. *JMP® 8 Introductory Guide, Second Edition*. Cary, NC: SAS Institute Inc.

JMP® 8 Introductory Guide, Second Edition

Copyright © 2009, SAS Institute Inc., Cary, NC, USA

ISBN 978-1-60764-299-2

SAS Institute Inc., SAS Campus Drive, Cary, North Carolina 27513.

1st printing, December 2009

Contents
JMP Introductory Guide

3   Summarizing Data
Look Closely at the Data

4   Looking at Distributions
Histograms, moments, quantiles, and proportions

5   Comparing Group Means
Testing Differences

6   Analyzing Categorical Data
Comparing Proportions

7   Regression and Curve Fitting
Visualizing Relationships

8   A Factorial Analysis
Designed Modeling

9 Exploring Data
Finding Exceptions

10 Multiple Regression
Examining Multiple Explanations

Index
JMP Introductory Guide

Credits and Acknowledgments

Origin

JMP was developed by SAS Institute Inc., Cary, NC. JMP is not a part of the SAS System, though portions of JMP were adapted from routines in the SAS System, particularly for linear algebra and probability calculations. Version 1 of JMP went into production in October, 1989.

Credits

JMP was conceived and started by John Sall. Design and development were done by John Sall, Chung-Wei Ng, Michael Hecht, Richard Potter, Brian Corcoran, Annie Dudley Zangi, Bradley Jones, Craige Hales, Chris Gotwalt, Paul Nelson, Xan Gregg, Jianfeng Ding, Eric Hill, John Schroedl, Laura Lancaster, Scott McQuiggan, and Peng Liu.

In the SAS Institute Technical Support division, Wendy Murphrey and Toby Trott provide technical support and conduct test site administration. Statistical technical support is provided by Craig DeVault, Duane Hayes, Elizabeth Edwards, Kathleen Kiernan, and Tonya Mauldin.

Nicole Jones, Jim Borek, Kyoko Keener, Hui Di, Joseph Morgan, Wenjun Bao, Fang Chen, Susan Shao, Hugh Crews, Yusuke Ono and Kelci Miclaus provide ongoing quality assurance. Additional testing and technical support is done by Noriki Inoue, Kyoko Takenaka, and Masakazu Okada from SAS Japan.

Bob Hickey is the release engineer.

The JMP manuals were written by Ann Lehman, Lee Creighton, John Sall, Bradley Jones, Erin Vang, Melanie Drake, Meredith Blackwelder, Diane Perhac, Jonathan Gatlin, and Susan Conaghan with contributions from Annie Dudley Zangi and Brian Corcoran. Creative services and production was done by SAS Publications. Melanie Drake implemented the help system.

Jon Weisz and Jeff Perkinson provided project management. Also thanks to Lou Valente, Ian Cox, Mark Bailey, and Malcolm Moore for technical advice.

Thanks also to Georges Guirguis, Warren Sarle, Gordon Johnston, Duane Hayes, Russell Wolfinger, Randall Tobias, Robert N. Rodriguez, Ying So, Warren Kuhfeld, George MacKensie, Bob Lucas, Warren Kuhfeld, Mike Leonard, and Padraic Neville for statistical R&D support. Thanks are also due to Doug Melzer, Bryan Wolfe, Vincent DelGobbo, Biff Beers, Russell Gonsalves, Mitchel Soltys, Dave Mackie, and Stephanie Smith, who helped us get started with SAS Foundation Services from JMP.

Acknowledgments

We owe special gratitude to the people that encouraged us to start JMP, to the alpha and beta testers of JMP, and to the reviewers of the documentation. In particular we thank Michael Benson, Howard Yetter (d), Andy Mauromoustakos, Al Best, Stan Young, Robert Muenchen, Lenore Herzenberg, Ramon Leon, Tom Lange, Homer Hegedus, Skip Weed, Michael Emptage, Pat Spagan, Paul Wenz, Mike Bowen, Lori Gates, Georgia Morgan, David Tanaka, Zoe Jewell, Sky Alibhai, David Coleman,

Linda Blazek, Michael Friendly, Joe Hockman, Frank Shen, J.H. Goodman, David Iklé, Barry Hembree, Dan Obermiller, Jeff Sweeney, Lynn Vanatta, and Kris Ghosh.

Also, we thank Dick DeVeaux, Gray McQuarrie, Robert Stine, George Fraction, Avigdor Cahaner, José Ramirez, Gudmunder Axelsson, Al Fulmer, Cary Tuckfield, Ron Thisted, Nancy McDermott, Veronica Czitrom, Tom Johnson, Cy Wegman, Paul Dwyer, DaRon Huffaker, Kevin Norwood, Mike Thompson, Jack Reese, Francois Mainville, and John Wass.

We also thank the following individuals for expert advice in their statistical specialties: R. Hocking and P. Spector for advice on effective hypotheses; Robert Mee for screening design generators; Roselinde Kessels for advice on choice experiments; Greg Piepel, Peter Goos, J. Stuart Hunter, Dennis Lin, Doug Montgomery, and Chris Nachtsheim for advice on design of experiments; Jason Hsu for advice on multiple comparisons methods (not all of which we were able to incorporate in JMP); Ralph O'Brien for advice on homogeneity of variance tests; Ralph O'Brien and S. Paul Wright for advice on statistical power; Keith Muller for advice in multivariate methods, Harry Martz, Wayne Nelson, Ramon Leon, Dave Trindade, Paul Tobias, and William Q. Meeker for advice on reliability plots; Lijian Yang and J.S. Marron for bivariate smoothing design; George Milliken and Yurii Bulavski for development of mixed models; Will Potts and Cathy Maahs-Fladung for data mining; Clay Thompson for advice on contour plotting algorithms; and Tom Little, Damon Stoddard, Blanton Godfrey, Tim Clapp, and Joe Ficalora for advice in the area of Six Sigma; and Josef Schmee and Alan Bowman for advice on simulation and tolerance design.

For sample data, thanks to Patrice Strahle for Pareto examples, the Texas air control board for the pollution data, and David Coleman for the pollen (eureka) data.

Translations

Erin Vang, Trish O'Grady, Elly Sato, and Kyoko Keener coordinate localization. Special thanks to Noriki Inoue, Kyoko Takenaka, Masakazu Okada, Naohiro Masukawa and Yusuke Ono (SAS Japan); and Professor Toshiro Haga (retired, Tokyo University of Science) and Professor Hirohiko Asano (Tokyo Metropolitan University) for reviewing our Japanese translation; Professors Fengshan Bai, Xuan Lu, and Jianguo Li at Tsinghua University in Beijing, and their assistants Rui Guo, Shan Jiang, Zhicheng Wan, and Qiang Zhao; and William Zhou (SAS China) and Zhongguo Zheng, professor at Peking University, for reviewing the Simplified Chinese translation; Jacques Goupy (consultant, ReConFor) and Olivier Nuñez (professor, Universidad Carlos III de Madrid) for reviewing the French translation; Dr. Byung Chun Kim (professor, Korea Advanced Institute of Science and Technology) and Duk-Hyun Ko (SAS Korea) for reviewing the Korean translation; Bertram Schäfer and David Meintrup (consultants, StatCon) for reviewing the German translation; Patrizia Omodei, Maria Scaccabarozzi, and Letizia Bazzani (SAS Italy) for reviewing the Italian translation. Finally, thanks to all the members of our outstanding translation teams.

Past Support

Many people were important in the evolution of JMP. Special thanks to David DeLong, Mary Cole, Kristin Nauta, Aaron Walker, Ike Walker, Eric Gjertsen, Dave Tilley, Ruth Lee, Annette Sanders, Tim Christensen, Jeff Polzin, Eric Wasserman, Charles Soper, Wenjie Bao, and Junji Kishimoto. Thanks to SAS Institute quality assurance by Jeanne Martin, Fouad Younan, and Frank Lassiter. Additional testing for Versions 3 and 4 was done by Li Yang, Brenda Sun, Katrina Hauser, and Andrea Ritter.

Also thanks to Jenny Kendall, John Hansen, Eddie Routten, David Schlotzhauer, and James Mulherin. Thanks to Steve Shack, Greg Weier, and Maura Stokes for testing JMP Version 1.

Thanks for support from Charles Shipp, Harold Gugel (d), Jim Winters, Matthew Lay, Tim Rey, Rubin Gabriel, Brian Ruff, William Lisowski, David Morganstein, Tom Esposito, Susan West, Chris Fehily, Dan Chilko, Jim Shook, Ken Bodner, Rick Blahunka, Dana C. Aultman, and William Fehlner.

Technology License Notices

The ImageMan DLL is used with permission of Data Techniques, Inc.

Scintilla is Copyright 1998-2003 by Neil Hodgson <neilh@scintilla.org>. NEIL HODGSON DISCLAIMS ALL WARRANTIES WITH REGARD TO THIS SOFTWARE, INCLUDING ALL IMPLIED WARRANTIES OF MERCHANTABILITY AND FITNESS, IN NO EVENT SHALL NEIL HODGSON BE LIABLE FOR ANY SPECIAL, INDIRECT OR CONSEQUENTIAL DAMAGES OR ANY DAMAGES WHATSOEVER RESULTING FROM LOSS OF USE, DATA OR PROFITS, WHETHER IN AN ACTION OF CONTRACT, NEGLIGENCE OR OTHER TORTIOUS ACTION, ARISING OUT OF OR IN CONNECTION WITH THE USE OR PERFORMANCE OF THIS SOFTWARE.

XRender is Copyright © 2002 Keith Packard. KEITH PACKARD DISCLAIMS ALL WARRANTIES WITH REGARD TO THIS SOFTWARE, INCLUDING ALL IMPLIED WARRANTIES OF MERCHANTABILITY AND FITNESS, IN NO EVENT SHALL KEITH PACKARD BE LIABLE FOR ANY SPECIAL, INDIRECT OR CONSEQUENTIAL DAMAGES OR ANY DAMAGES WHATSOEVER RESULTING FROM LOSS OF USE, DATA OR PROFITS, WHETHER IN AN ACTION OF CONTRACT, NEGLIGENCE OR OTHER TORTIOUS ACTION, ARISING OUT OF OR IN CONNECTION WITH THE USE OR PERFORMANCE OF THIS SOFTWARE.

Chapter **1**

Introducing JMP
Your First Look

JMP uses an extraordinary graphical interface to display and analyze data. JMP is software for interactive statistical graphics and includes:

- a data table window for editing, entering, and manipulating data
- a broad range of graphical and statistical methods for data analysis
- an extensive design of experiments module
- options to highlight and display subsets of data
- a formula editor for each table column to compute values as needed
- a facility for grouping data and computing summary statistics
- special plots, charts, and communication capability for quality improvement techniques
- tools for printing and for moving analyses results between applications
- a scripting language for saving and creating frequently used routines

This introductory chapter gives basic information about using JMP.

Contents

What You Need to Know

Before you begin using JMP, you should be familiar with:

- Standard operations and terminology such as *click*, *double-click*, *Ctrl-click*, and *Alt-click* (*Command-click* and *Option-click* on the Macintosh), *Shift-click*, *drag*, *select*, *copy*, and *paste*.

- How to use menu bars and scroll bars, how to move and resize windows, and how to manipulate files in the desktop. If you are using your computer for the first time, consult the reference guides that came with it for more information.

- Minimal statistics. Even though JMP has many advanced features, you only need a minimal background of formal statistical training. All analyses include graphical displays with options that help you review and interpret the results. Each analysis also includes access to help windows that offer general help and some statistical details.

Learning About JMP

If you are familiar with JMP, you might want to know only what's new. The *JMP New Features* document gives a summary of general changes and additions. To learn more about JMP, use the recommendations in the following sections.

Using Tutorials

JMP provides three types of tutorials:

- **Beginner's Tutorial** The beginner's tutorial steps you through the JMP interface and explains the basics of how to use JMP. It is accessible through the Tip of the Day window, which appears when you start JMP. To start the tutorial from the Tip of the Day window, click **Enter Beginner's Tutorial**. Or, start the tutorial by selecting **Help** (**View** on the Macintosh) **> Tutorials > Beginners Tutorial**.

- **Specific Analysis Tutorials** Tutorials that step you through creating an analysis in JMP are found under **Help** (**View** on the Macintosh) **> Tutorials**. Tutorials describe how to create a chart, compare means, how to design an experiment, and more.

- **JMP Introductory Guide** The *JMP Introductory Guide* is a collection of tutorials designed to help you learn JMP strategies. If you did not receive a printed copy of this book, view the .pdf file by selecting **Help > Books > JMP Introductory Guide**. By following along with these step-by-step examples, you can quickly become familiar with JMP menus, options, and report windows.

Searching in the Help

You might want help on a specific topic, and you want to search the online Help for that topic. The main menu bar contains a Help menu, which provides the appropriate searching capabilities.

On Windows and Linux, the **Help > Contents**, **Help > Search**, and **Help > Index** commands access the JMP Help system. The Help system provides navigable online JMP documentation.

On the Macintosh, the **Help > JMP Help** command displays a list of JMP help items with search capabilities and a table of contents.

Learning About Statistical and JSL Terms

The **Help > Indexes** command displays the following sources for your reference:

- **Statistics Index**   Accesses references that give definitions of statistical terms. Once you are in the Statistics Index window, click **Topic Help** to go to the place in the online Help that describes the highlighted topic. Click **Example** to run the script associated with the highlighted topic. Click **Launch** to run the script that corresponds to the item you have highlighted in the list.

Figure 1.1 The Statistics Index

    A List of topics
    B Description
    C Example script

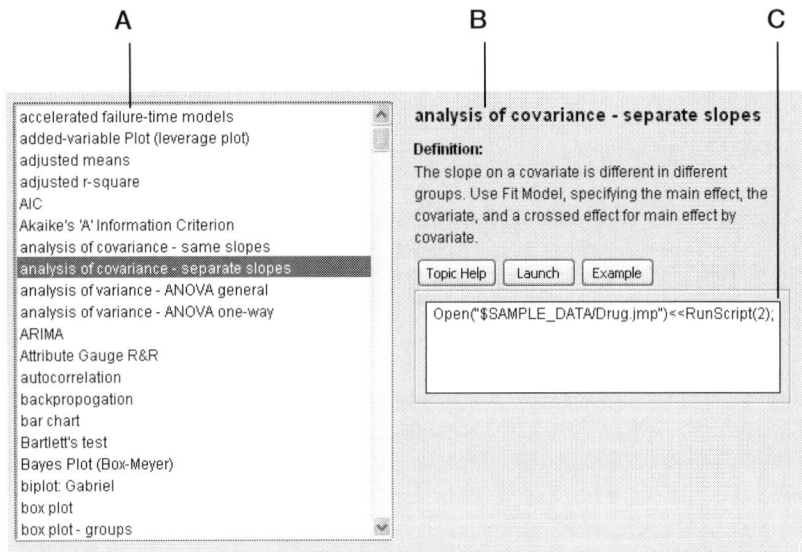

- **JSL Functions Index**   Presents a list of JSL operators, such as Sin, Cos, Sqrt, and Abbrev Date that you would use when writing JSL. Highlight an operator name to see a description of the operator appear in the window on the right. Click **Topic Help** to see more information in the online Help.

- **Object Scripting Index**   Presents a list of JSL objects. These are scriptable JSL building blocks. Highlight an object name and messages the object recognizes appear in the window on the right.

- **DisplayBox Scripting Index**   Presents a list of the elements that make up a JMP report. These elements are the JSL building blocks with which you build output. Highlight a Display Box and available messages for each object appear in the window on the right.

Using the Context-Sensitive Help

To use the online Help system, select one of the following methods:

- Select **Help** from analysis construction windows (as shown in Figure 1.2) and report windows.

Figure 1.2 Help Is Available

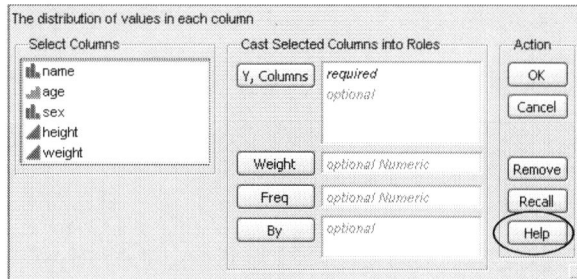

- Select the help tool (?) from the **Tools** menu and click a place in a data table or report on which you need assistance (Figure 1.3). Context-sensitive help tells about the items in the area you clicked.

Figure 1.3 Use the Help Tool for Context-Sensitive Help

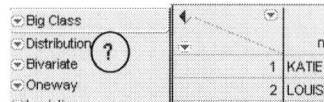

- In some reports, make a small circle with your cursor to reveal information about the item in the area.

Figure 1.4 Making a Circle with the Cursor Displays Help

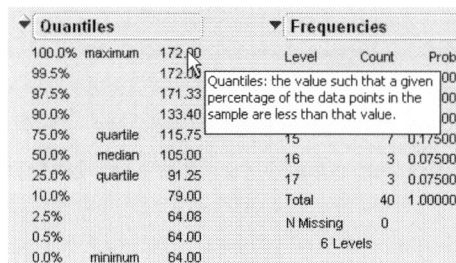

- In some menus, hold the cursor on menu items to reveal information about the menu item.

Figure 1.5 Display a Description of Menu Items

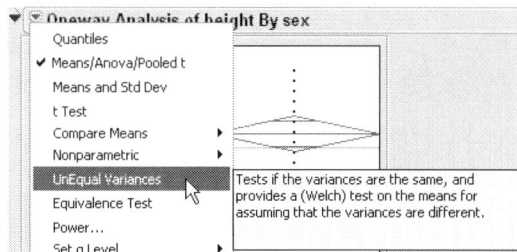

Learning JMP Tips & Tricks

When you first start JMP, you see the Tip of the Day window. This window provides tips about using JMP that you might not know.

To turn off the Tip of the Day, clear the **Show tips at startup** check box. To view it again, select **Help** (**View** on the Macintosh) **> Tip of the Day**.

Also use the *JMP Quick Reference Guide* to learn more advanced commands in JMP. View this document by selecting **Help > Books > JMP Quick Reference Card**.

Using This Book in Combination with Other Included Books

The book you are reading now is the *JMP Introductory Guide*. See the following manuals for further documentation of JMP:

- The *JMP User Guide* has complete documentation of all JMP menus, an explanation of data table manipulation, and a description of the formula editor. There are chapters that show how to do common tasks such as manipulating files, transforming data table columns, and cutting and pasting JMP data, statistical text reports, and graphical displays.

- The *JMP Statistics and Graphics Guide* gives documentation of the **Analyze** and **Graph** menus. It documents analyses, discusses statistical methods, and describes all report windows and options.

- The *JMP Design of Experiments* covers the **DOE** menu, the experimental design facility in JMP.

- The *JMP Scripting Guide* is a reference guide to the JMP scripting language (JSL) that lets you automate action sequences.

If you did not receive printed copies of these books, view the .pdf files by selecting **Help > Books**.

Conventions Used in this Book

Conventions used in this manual were devised to help relate written material to information that appears on-screen:

- The .jmp extension follows filenames on the PC. When you installed JMP, a folder named Sample Data was also installed. On the Macintosh, JMP sample data files have the same name, but show without an extension. Reference to names of JMP files, data tables, variable names, and items in reports appear in Helvetica to help distinguish them from surrounding text.

- Special information, warnings, and limitations are noted in sentences beginning with the bold word **Note**.

- Reference to menu names (**File** menu) or menu items (**Save** command) appear in **Helvetica bold** font.

- The notation to select a command from a menu is sometimes written as **File > New,** meaning "select the **New** command from the **File** menu."

- Words or phrases that are important or have definitions specific to JMP are in *italics* the first time they appear.

Step 1: Start JMP

Start a JMP session by double-clicking the JMP application icon. Your initial view of JMP is a menu bar, a toolbar, the Tip of the Day window, and the JMP Starter window (Figure 1.6).

Figure 1.6 First View of JMP (Windows)

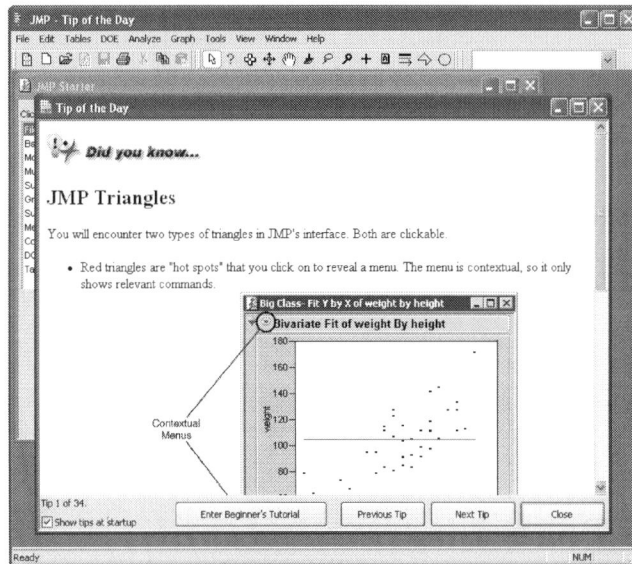

To start the online tutorial, click **Enter Beginner's Tutorial**. Or, click the **Close** button to close the window and follow the tutorials in this book.

Step 2: Open a JMP Data Table

There are several ways to open a data table:

- Go to your Sample Data directory in its default location. For example:

  In Windows: C:\Program Files\SAS\JMP\8\Support Files English\Sample Data

  In Linux: /Opt/SAS/JMP8/Support Files English/Sample Data

  In Macintosh: /Library/Application Support/JMP/8/English/Sample Data

- Selecting **File > New** (or clicking the **New Data Table** button on the JMP Starter window) creates and displays a data table with an empty data grid. First, add rows and columns, then type in or paste in new data. For details, see the *JMP User Guide*.

- Selecting **File > Open** (or clicking the **Open Data Table** button on the JMP Starter window) presents a file selection window (Figure 1.7) with a list of existing tables. Select a file and click **Open**. For details, see the *JMP User Guide*.

Figure 1.7 The Open Data File Window

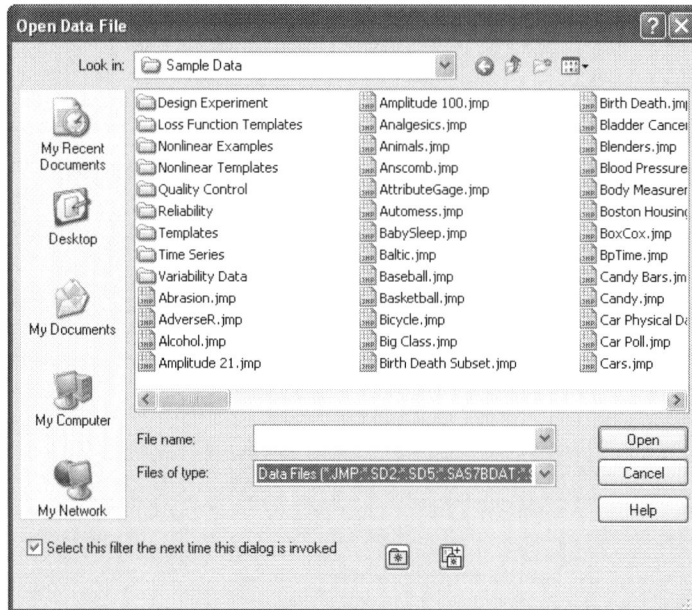

Step 3: Learn About the Data Table

Opening or creating a data table creates a data grid and table information panels, like the ones shown in Figure 1.8. The counts of table rows and columns appear in the corresponding panels to the left of the data grid. In the data grid, a row number identifies each row, and each column has a column name. Rows and columns are sometimes called *observations* and *variables* in an analysis.

Figure 1.8 A Data Table

The JMP data table window is a flexible way to prepare data. Using it, you can accomplish a variety of table management tasks, such as:

- Editing the value in any cell

- Changing a column's width by dragging the column line
- Hiding columns temporarily, or deleting columns permanently
- Adding rows, or rearranging the order of rows
- Adding columns, or rearranging the order of columns
- Selecting a subset of rows for analysis and saving that subset for further use
- Sorting or combining tables

For details, see the *JMP User Guide*.

Specifying the Values' Type

The small icon to the left of the column name in the columns panel is an icon that can be clicked. Use it to declare the modeling type of the values in the column. JMP uses three modeling types to determine how to analyze the column's values:

Continuous ( )   Values are numeric measurements.

Ordinal ( )   Values are ordered categories, which can have either numeric or character values.

Nominal ( )   Values are numeric or character classifications.

Modeling types are changeable depending on how you want to look at your data. For example, a variable like age should be specified continuous to find the mean (average) age, but nominal or ordinal to find frequency counts for each age value.

The default modeling type is nominal for character values and continuous for numeric values.

To assign a different modeling type to a variable:

1   Click the icon next to the variable name.

2   Select the appropriate modeling type.

Figure 1.9 Changing a column's modeling type

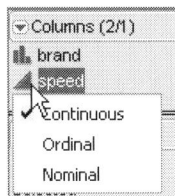

For details, see the *JMP User Guide*.

Data Table Cursor Forms

As you move the cursor around the data table, it changes forms. Its shape gives you information about performable actions. The following sections describe the different cursor forms.

Arrow Cursor

The cursor is a standard arrow when it is anywhere in the table panels to the left of the data grid, except when it is on a red triangle icon (⊛) or diamond-shaped disclosure button (◆ ◆ on Windows and ▶ ▼ on the Macintosh) or when it is in the upper-left corner of the data grid—the area where rows and columns are *de*selected. (See Figure 1.10.)

Figure 1.10

    A Click to select a column. Double-click to edit a column name.
    B Click in this area to deselect all rows.
    C Click in this area to deselect all columns.

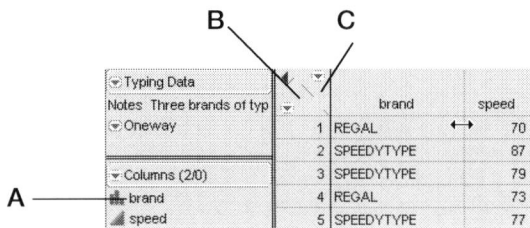

I-Beam Cursor

The cursor is an I-beam when it is over text in the data grid or highlighted column names in the data grid or column panels. To edit text in the data grid:

1 Click the cell that you want to edit. The cell highlights.

2 Click again next to any character to mark an insertion point.

3 The I-beam deposits a vertical blinking bar.

4 Use the keyboard to make changes.

To edit a column name:

1 Click the column name to highlight the column.

2 Press the Enter key to change the I-beam cursor to an insertion point.

3 Use the keyboard to make changes.

Large Cross Cursor

The cursor becomes a large cross when moved into a column or row selection area. When moved over a column name, you can edit the name. To do so, click the column name and begin typing.

The cross cursor can also be used to select rows and columns. To select a column, click the area above the column name. See the next section, "Selecting Rows and Columns," p. 11, for a detailed explanation of selecting rows and columns.

Double Arrow Cursor

The cursor changes to a double-arrow cursor when positioned on a column boundary or on a panel splitter. Dragging the double-arrow cursor changes the column width or the panel size.

Figure 1.11 Changing the width of a column

   A Click and drag to change the width of a column.

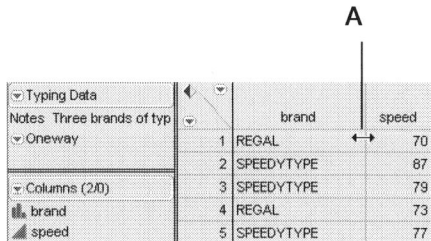

☝ Hand Cursor

The cursor changes to a hand when you move the mouse over a red triangle icon (☉) or diamond-shaped disclosure button (◀ ◆ on Windows and ▶ ▼ on the Macintosh).

Click the red triangle to reveal the menu and select a menu icon. Click the disclosure button to open or close a panel.

Figure 1.12 The hand cursor

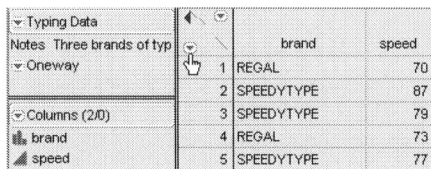

Selecting Rows and Columns

Select rows and columns in a JMP data grid by highlighting them, as explained in Table 1.1 and shown in Figure 1.13. For additional details, see the *JMP User Guide*.

Table 1.1 Ways to Select Rows and Columns

Action	Instructions
Highlight a row	Click the space that contains the row number.
Highlight a column	Click the background area above the column name. Or, click the column name in the columns panels to the left of the data grid.
Extend a selection of rows or columns	Shift-click the first and last rows or columns of the desired range.
Make a discontiguous selection	Ctrl-click (Command-click on the Macintosh) the desired selections.

Figure 1.13 Select Rows and Columns

A Selected rows
B Selected column

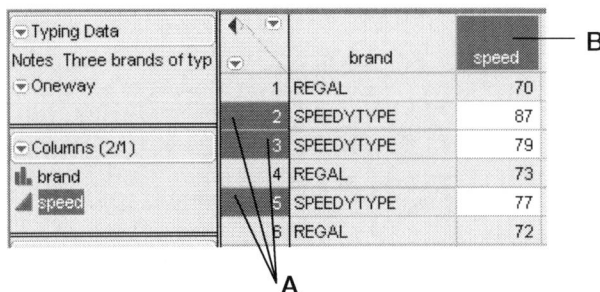

Step 4: Select an Analysis

There are a variety of analyses available through the **Analyze** and **Graph** menus in the main menu. An alternate way to access these analyses is through toolbar buttons and selections in the JMP Starter window. Selecting an analysis in the **Analyze** or **Graph** menus produces graphs, charts, plots, and tables. For example, to see a histogram of columns in the data table you have open, select **Analyze > Distribution**. Then, complete the window and click **OK**.

Casting Columns Into Roles

After you select an analysis from the main menu, a window appears that asks you to cast columns into roles. For example, if you select **Analyze > Fit Y by X** from the main menu, you see the window in Figure 1.14.

Figure 1.14 Fit Y by X Window

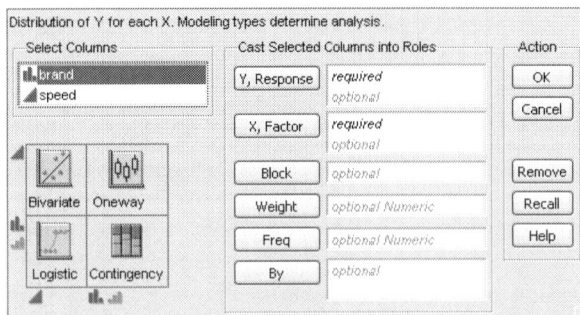

The JMP analysis methods are like stages or platforms for variables to dramatize their values. Each analysis requires information about which variables play what roles in an analysis.

The most typical variable roles are:

- **Y, Response**   Identifies a column as a response or dependent variable whose distribution is to be studied.
- **X, Factor**   Identifies a column as an independent, classification, or explanatory variable whose values divide the rows into sample groups.
- **Weight**   Identifies a numeric column whose values supply weights for each response.
- **Freq**   Identifies a numeric column whose values assign a frequency to each row for the analysis.
- **By**   Identifies a column that is used to create a report consisting of separate analyses.

Step 5: View the Output Report

After you have cast columns into their roles, JMP provides output reports that include graphics and text. For more detail than is presented below, see the *JMP User Guide*.

Graphs and Charts

JMP reports are usually filled with graphs, charts, plots, and other graphical displays that show your results. For example, if you select **Analyze > Distribution** and assign several columns the **Y, Response** role in the Distribution window that appears, you create a report that contains a graphical display of each column assigned the **Y, Response** role.

For the example shown in Figure 1.15, the **Distribution** command produces graphical displays that include:

- Histograms of both the brand and speed columns.
- An outlier box plot of the continuous variable speed.

Figure 1.15 Distribution Histograms and Outlier Box Plot

A   Histogram
B   Outlier Box Plot

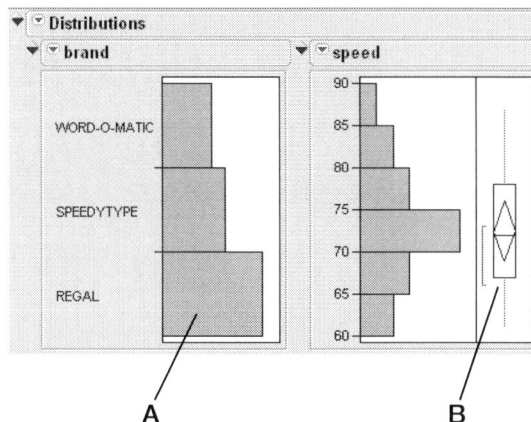

Display Options

To enhance the default graphical displays that show your results, JMP provides options that you can add to them. These options are found by clicking the red triangle icon beside a report name. For example, the red triangle icon next to the histogram name lists available report options (Figure 1.16). For practice, try selecting different combinations of these options and watch the effect they have on the displays and reports.

🖰 Select a column in Select Columns.

🖰 Select **Y, Columns** in Cast Selected Columns into Roles.

🖰 Select variables for **Weight**, **Freq**, and **By**.

🖰 Click **OK**.

Figure 1.16 Options for Nominal or Ordinal Variable in a Distribution Analysis

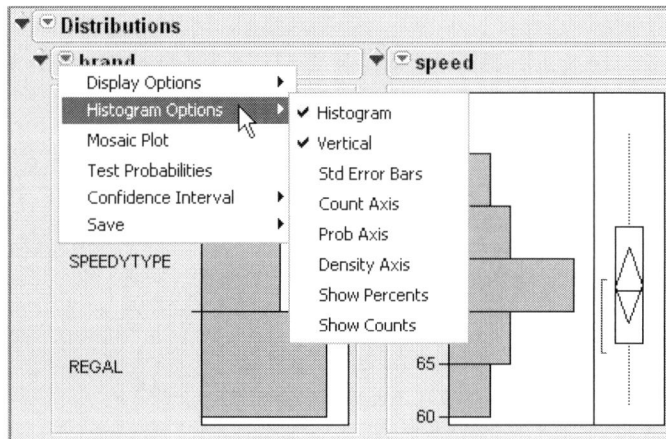

Statistical Tables and Text

In addition to the graphs, charts, plots, and other graphical displays in a report, JMP can also include text tables in a report. The types of tables given depend on whether a variable is continuous or categorical (ordinal or nominal).

When you installed JMP, a folder named **Sample Data** was also installed. In that folder is a file named **Typing Data.jmp**. Using the **Distribution** command to produce a distribution report creates text tables for the speed and brand variables (Figure 1.17). In this example:

- For a continuous variable, JMP displays a Quantiles table and a Moments table.
- For nominal and ordinal variables, JMP displays a Frequency table showing the total sample frequency, category frequencies, and associated probabilities.

Figure 1.17 Statistical Reports

Distributions

brand

Frequencies

Level	Count	Prob
REGAL	8	0.47059
SPEEDYTYPE	5	0.29412
WORD-O-MATIC	4	0.23529
Total	17	1.00000

N Missing 0
3 Levels

speed

Quantiles

100.0%	maximum	87.000
99.5%		87.000
97.5%		87.000
90.0%		82.200
75.0%	quartile	78.000
50.0%	median	72.000
25.0%	quartile	67.000
10.0%		61.800
2.5%		61.000
0.5%		61.000
0.0%	minimum	61.000

Moments

Mean	72.470588
Std Dev	7.0010503
Std Err Mean	1.6980041
upper 95% Mean	76.070196
lower 95% Mean	68.87098
N	17

JMP also gives you the ability to change the appearance of these tables. For details, see the *JMP User Guide*.

Step 6: Save the JMP Output Report

To save a report just as it appears in the report window, select **File > Save**.

Another way to save is to duplicate it in a separate window titled Journal. Then you can append other reports to it or manipulate it in the journal window. To do this, select **Edit > Journal.** Then, select **File > Save As**. For more detail than is presented below, see the *JMP User Guide*.

Figure 1.18 A Journal

A The report window

B The journal window

A

B

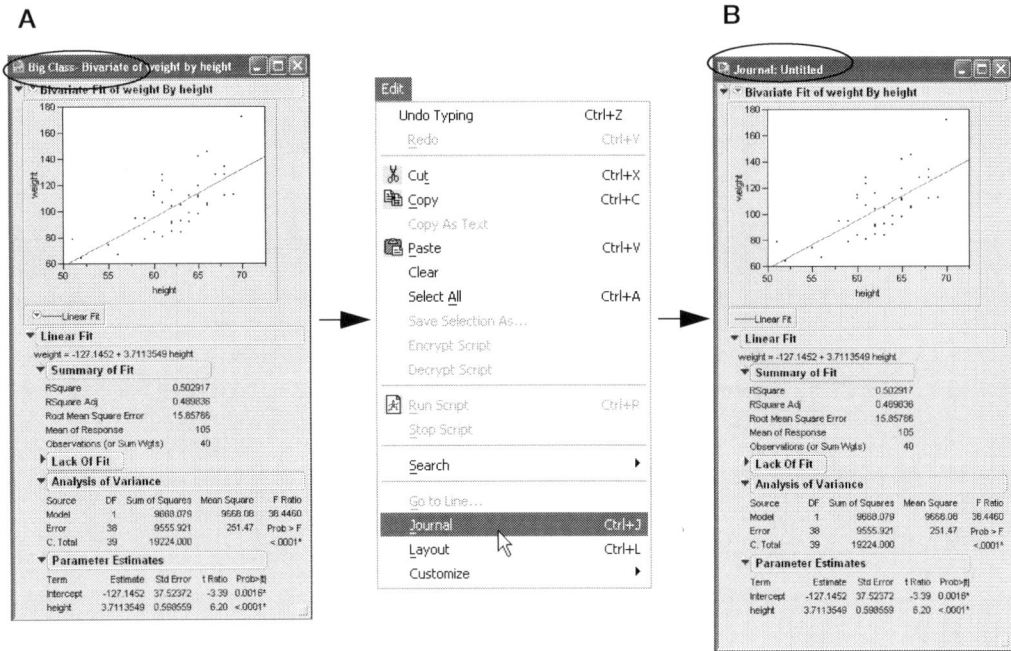

A Practice Tutorial

Before you begin the tutorials in the following chapters of this book, complete this brief practice tutorial that is a short guided tour through a JMP analysis. Follow the steps to see a three-dimensional scatterplot.

Open a Data Table

⌐ Open the file called **Cowboy Hat.jmp** to begin a JMP session. When you installed JMP, a folder named **Sample Data** was also installed. In that folder is a file named **Cowboy Hat.jmp**.

The data table shown in Figure 1.19 appears.

Figure 1.19 Cowboy Hat Data Table

	x	y	z	hue	hue, shade
1	-5	-5	0.70886129		
2	-5	-4.5	0.42921793		
3	-5	-4	0.11965158		
4	-5	-3.5	-0.1789386		
5	-5	-3	-0.4369755		
6	-5	-2.5	-0.6388599		
7	-5	-2	-0.7820949		
8	-5	-1.5	-0.8738338		
9	-5	-1	-0.9261848		
10	-5	-0.5	-0.9515529		

Cowboy Hat — Notes The values for x, y,

Columns (5/0) — x, y, z, hue, hue, shade

Rows — All rows 401 — Selected 0

This data table has three numeric columns and two *row state* columns. Columns x and y are *x*- and *y*-coordinates, and z is created using the function

$$z = \sin\sqrt{x^2 + y^2}$$

Select an Analysis

To plot the three columns of information from the **Cowboy Hat** data table:

🖰 Choose the **Scatterplot 3D** command from the **Graph** menu.

🖰 Select the x, y, and z columns from the column selector list on the left side of the window, and click **Y, Columns**, as shown in Figure 1.20.

Figure 1.20 Scatterplot 3D Column Selection Window

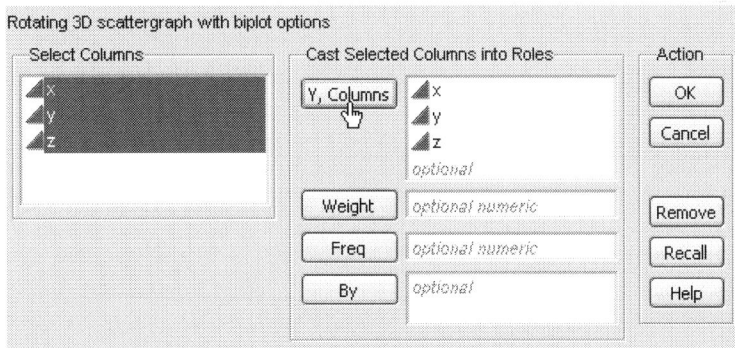

Rotating 3D scattergraph with biplot options

Select Columns: x, y, z

Cast Selected Columns into Roles:
- Y, Columns: x, y, z — optional
- Weight: optional numeric
- Freq: optional numeric
- By: optional

Action: OK, Cancel, Remove, Recall, Help

These column names now appear in the list on the right side of the window.

🖰 Click **OK**.

The scatterplot 3-D appears. (See Figure 1.21.)

Figure 1.21 The Cowboy Hat

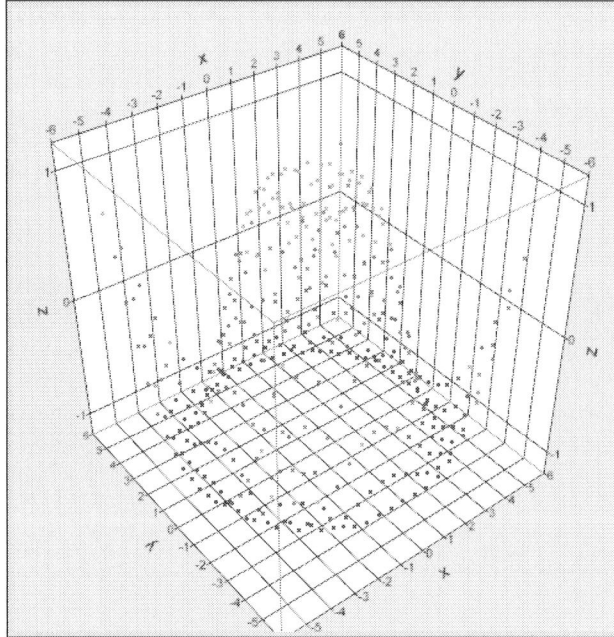

Spin the Cowboy Hat

To get a better view of the cowboy hat,

🖱 Right-click on the plot and select **Settings**.

🖱 Deselect **Walls**, **Grids**, **Axes**, and **Box**.

🖱 Move the mouse over the plot, press the mouse button, and move the cursor about.

The cowboy hat moves in three dimensions.

Figure 1.22 Moving the cowboy hat

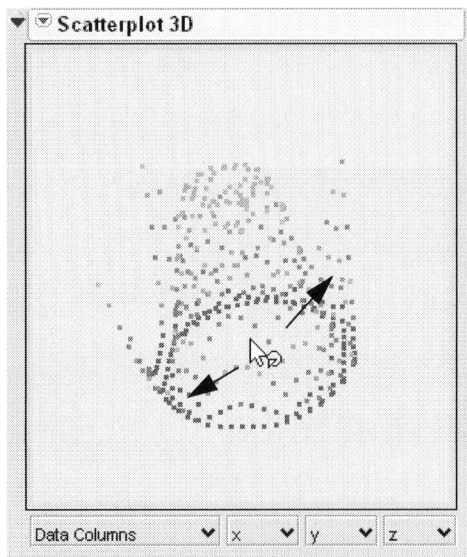

To have the plot spin itself:

🖑 Press the Shift key, and give the plot a push with the cursor.

🖑 To stop the spinning, click again in the scatterplot 3-D frame.

Chapter **2**

Creating a JMP Data Table
Entering and Plotting Data

This lesson evaluates a new drug developed to lower blood pressure. Data were recorded over a six-month period for the following treatment groups:

- 300 mg dose
- 450 mg dose
- placebo
- control

Figure 2.1 shows the mean monthly blood pressure for each group, recorded in a journal. This lesson shows how to enter data values into the data table and to create a single neat and informative line plot that shows the study results.

Objectives

- Create rows and columns in a data table, one at a time and in groups.
- Enter data into JMP.
- Create a chart using the **Chart** command.
- Rescale axes in a plot.
- Animate a plot.

Figure 2.1 Blood Pressure Study

Blood Pressure Study

Month	Control	Placebo	300mg	450mg
March	165	163	166	168
April	162	159	165	163
May	164	158	161	153
June	162	161		
July	166	158		
August	163	158		

	Month	Control	Placebo	300mg	450mg
1	March	165	163	166	168
2	April	162	159	165	163
3	May	164	158	161	153
4	June	162	161	158	151
5	July	166	158	160	148
6	August	163	158	157	150

Contents

Starting a JMP Session

⍁ Open the JMP application icon to begin a JMP session.

⍁ Use the **New** command in the **File** menu to create an empty data table like the one shown here. Click **File > New > Data Table** or open JMP Starter and select **New Data Table**.

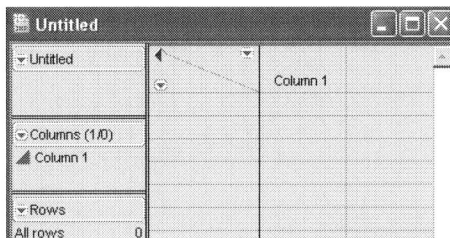

The data values for this project are blood pressure statistics collected over six months and recorded in a notebook page as shown in Figure 2.2.

Figure 2.2 Notebook of Raw Study Data

Blood Pressure Study

Month	Control	Placebo	300mg	450mg
March	165	163	166	168
April	162	159	165	163
May	164	158	161	153
June	162	161	158	151
July	166	158	160	148
August	163	158	157	150

Creating Rows and Columns in a JMP Data Table

JMP data tables have rows and columns, sometimes called *observations* and *variables* in statistical terms. The raw data in Figure 2.2 are arranged as five columns (treatment groups) and six rows (months March through August). The first line in the notebook names each column of values. These names can be used as column names in the new JMP table.

Add Columns

First create the number of rows and columns that are needed.

⍤ Select **Cols > Add Multiple Columns**, which prompts for the number of columns to add, where to add them, and which type of columns to add.

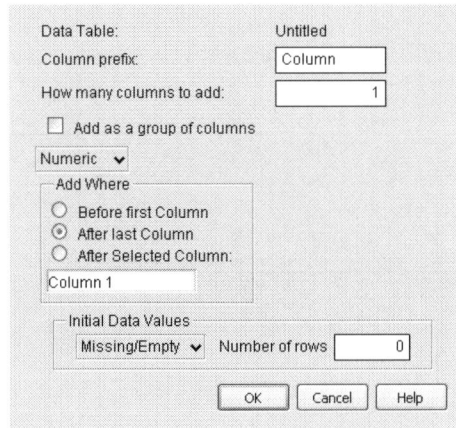

⍤ Specify five new columns.

The default column names are Column 1, Column 2, and so on. These can be changed at any time in the columns panel or at the top of the column in the data table. So, the next step is to type in meaningful names. Use the names from the data journal—Month, Control, Placebo, 300 mg, and 450 mg.

⍤ Edit a column name by clicking it in the data grid or in the columns panel.

⍤ Begin typing once the name is highlighted.

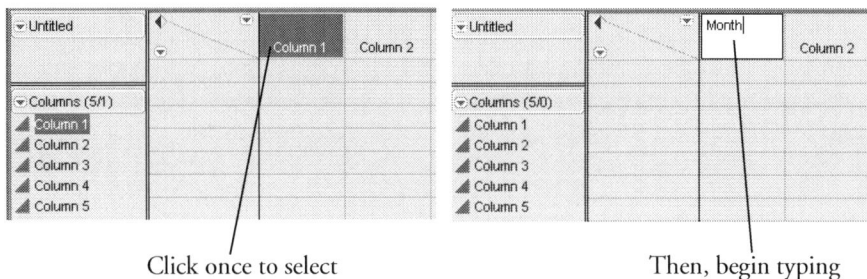

Click once to select   Then, begin typing

Set Column Characteristics

Columns can have different characteristics. By default, they contain numeric data. However, the values for month in this example are character values. To change the Month column from numeric to character:

⍤ Highlight the column by clicking either in the area at the top of the column or the area beside its name in the columns panel.

⍤ Select the **Cols > Column Info** to display the window in Figure 2.3.

⍤ Change Month to a character variable, as shown in Figure 2.3, by clicking the box beside **Data Type**.

The Column Info window is also used to change other column characteristics and to access the JMP formula editor for computing column values.

Figure 2.3 Change Data Type

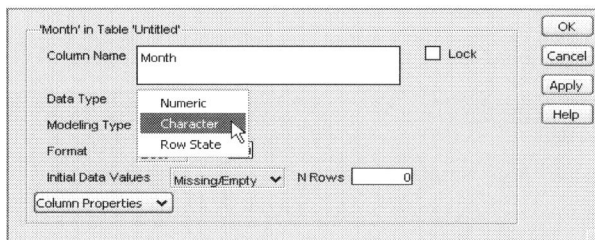

Add Rows

Now add new rows to the table:

🖑 Choose **Rows > Add Rows**.

🖑 Specify six new rows.

Alternatively, double-click anywhere in the body of the table to automatically fill it with new rows up through the position of the cursor.

🖑 Select **File > Save** to name the table **BP Study.jmp** and save it.

The data table is now ready to hold data values. To summarize the table evolution so far, you:

- Began with a new untitled table.
- Added enough rows and columns to accommodate the raw data.
- Tailored the characteristics of the table by giving the table and columns descriptive names.
- Changed the data type of the Month column to accept character values.

Entering Data

To enter data into the data table, type values into their appropriate table cells.

🖑 Type the values from the study journal (Figure 2.2) into the **BP Study.jmp** table as shown here.

When entering data into the data table:

- Edit the cell value by moving the cursor into a data cell and double-clicking. The cursor becomes a flashing vertical bar.
- Correct a mistake by dragging the text entry bar across the incorrect entry and typing the correction over it.
- Press the Return or Enter key on the *numeric keypad* or the Tab key to move the highlight one cell to the right. Press Shift-Tab or Shift-Return/Enter on the numeric keypad to move the highlight one cell to the left.

	Month	Control	Placebo	300mg	450mg
1	March	165	163	166	168
2	April	162	.	.	.
3			.	.	.	.
4			.	.	.	.
5			.	.	.	.
6			.	.	.	.

Click to highlight

	Month	Control	Placebo	300mg	450mg
1	March	165	163	166	168
2	April	162	159	.	.
3			.	.	.	.
4			.	.	.	.
5			.	.	.	.
6			.	.	.	.

Begin typing

The completed data table

	Month	Control	Placebo	300mg	450mg
1	March	165	163	166	168
2	April	162	159	165	163
3	May	164	158	161	153
4	June	162	161	158	151
5	July	166	158	160	148
6	August	163	158	157	150

Plotting Data

When working with the **Analyze** and **Graph** menu commands, you tell JMP which columns to work with and what to do with them. This section shows how to plot the months across the horizontal (x) axis and the columns of blood pressure statistics for each treatment group overlaid on the vertical (y) axis.

🖰 Select **Graph > Chart**.

The window in Figure 2.4 appears.

Assign x and y roles and choose the type of chart. This example specification is for a bar chart, with data (as opposed to statistics) as chart points.

🖰 Assure that the default choice **Vertical** is selected from the chart type drop-down list.

🖰 Select one continuous variable in the list.

🖰 Select the Shift and down arrow keys to select the other continuous variables.

🖰 Select the four continuous variables in the **Select Columns** list.

🖰 Click the **Statistics** button and select **Data** from the drop-down list.

🖰 Select **Month** from the **Select Columns** list.

🖰 Click **Categories**.

🖰 Click **OK**.

Figure 2.4 Creating the Bar Chart

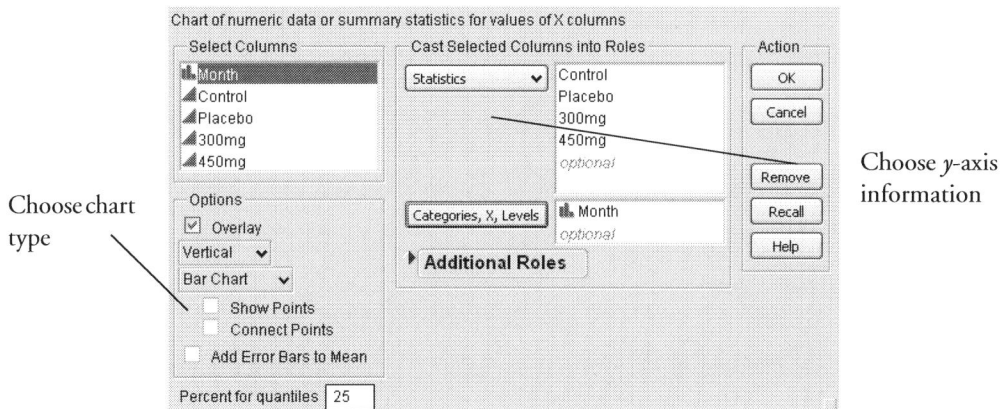

JMP displays an overlaid bar chart of the data.

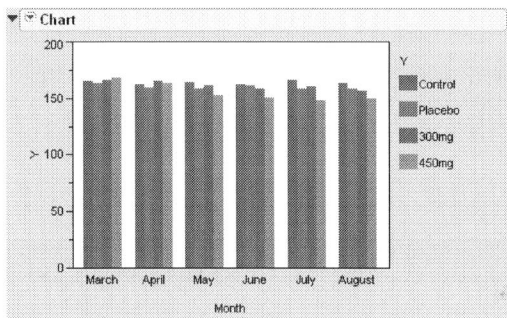

Reorder the Values

By default, the Month values in this chart do not appear in a logical order. To further enhance the report, reorder the months so they are in chronological order rather than alphabetical order.

🖰 In the data table, highlight the Month column and select **Cols > Column Info**.

🖰 Select **Value Ordering** from the **Column Properties** drop-down menu. Value ordering information appears on the right, as shown in Figure 2.5. The list in the window contains values in the order in which they appear in reports. Use the **Move Up** and **Move Down** buttons to change the order of the months.

Figure 2.5 The Value Ordering Window

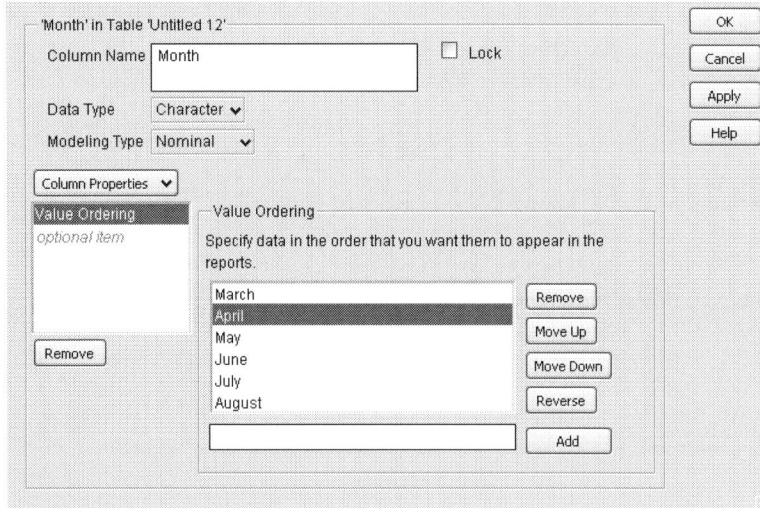

🖰 Click **OK**. The properties icon (✱) now appears next to the column name in the data table's column panel, indicating the column contains a property.

🖰 In the analysis report, click the red triangle and select **Script > Redo Analysis**.

Rescale the Plot Axis

By default, *y*-axis scaling begins at zero and the overlay chart looks like the one shown here. But, to present easy-to-read information, the *y*-axis needs to be rescaled and the chart needs labels.

🖰 Double-click the *y*-axis area, which accesses the Axis Specification window (Figure 2.6).

This window gives you the ability to:

- Set the minimum and maximum of the axis scale.
- Specify the tick mark increment.
- Request minor tick marks.
- Request grid lines at major or minor tick marks.
- Format numeric axes.
- Use either a linear or log-based scale.

Figure 2.6 Axis Specification Window

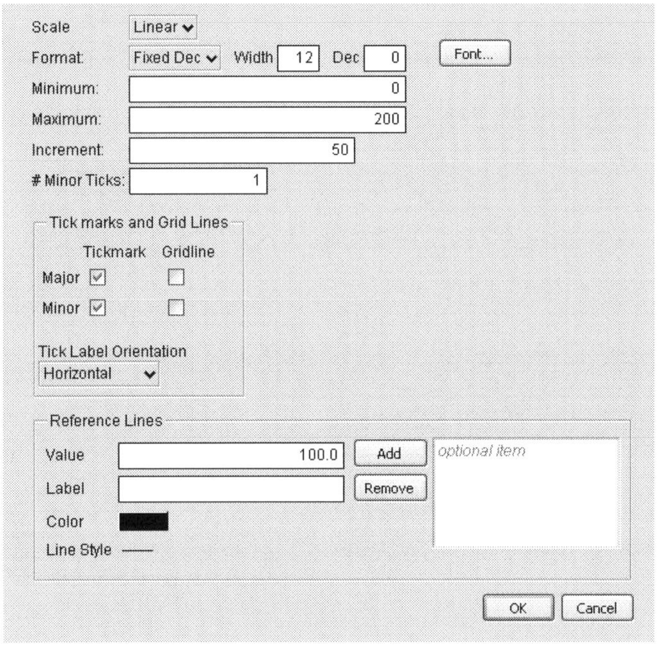

In this example, the plotted values range from about **145** to **175**.

🖑 Enter these figures into the Axis Specification window for **Minimum** and **Maximum**.

🖑 Change the increment for the tick marks from 50 to 1 by entering a **1** in the **Increment** box.

🖑 Click **OK**.

Tip: The magnifier tool (🔍), found in the **Tools** menu and the cursor toolbar, can also be used to change the scale of graphs. Drag the magnifier diagonally across the points of interest to see the chart automatically adjust. Double-click the plot frame to reset the plot to its original scale.

🖑 Click the edge of the graph and drag it to the right to increase its width.

🖑 Change the name of the axis from Y to Blood Pressure:

🖑 Place cursor over Analysis Report until cursor becomes an I-bar.

🖑 Click for a text box and enter **Blood Pressure**.

Document the Report

The chart also needs a title and other documentation to make it easy to interpret. The annotate tool (🄰) places text on the report. Refer to Figure 2.7 to see where to place the following steps.

🖑 Select the annotate tool.

🖑 Click and drag in the report to create a text box.

🖑 Release the button and enter the text for the title Comparison of Treatment Groups.

🖑 Click outside the annotation to quit editing the text.

🖑 Repeat to enter the footnote XYZ Blood Pressure Study 2007.

Note: Double-click any report title bar to edit the text on the bar.

Figure 2.7 Bar Chart with Modified Y-Axis, Titles, and Footnotes

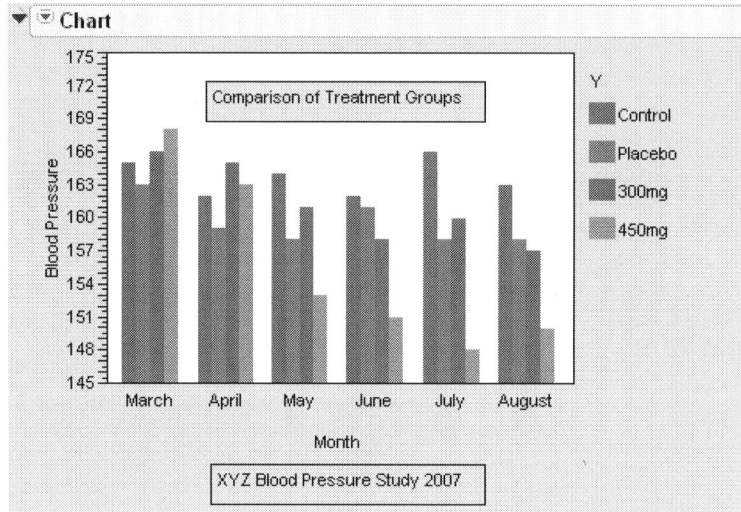

Chapter Summary

A study was done to evaluate the effect of a new drug on blood pressure. To complete this analysis, you:

- Used the **New Data Table** command in the **File** menu to create a new JMP table.
- Created the appropriate number of rows and columns for the data.
- Typed the data into the empty data grid.
- Used the **Chart** command in the **Graph** menu to request a bar chart of blood pressure measures over time.
- Ordered the values in chronological order so they would appear properly in the chart.
- Tailored the chart with a specific axis scale and axis name, and added a plot title and footnote with the annotate tool.

Chapter **3**

Summarizing Data
Look Closely at the Data

The hot dog is a questionable item on a school cafeteria menu because of its reputation as an unhealthy food, possibly classified in the junk food category. Many students feel this is unpatriotic and are upset.

This lesson examines the hot dog as a menu item, but not before looking into the multitude of brands available. The data shows information about cost, nutritional ingredients of concern, and taste preference for 54 hot dog brands. This information is sufficient to provide a summary of hot dog statistics and to identify the brands that are:

- most nutritious
- least costly
- best tasting

The taste, cost, and nutritional variables used in this chapter are an enhancement of data from Moore, D. S., and McCabe G. P., (1989), *Introduction to the Practice of Statistics*, and *Consumer Reports* (1986). The brand names were changed to fictional names, and the taste preference labels correspond to a taste preference scale.

Objectives

- Find and mark subgroups of data
- Produce scatterplots using the **Fit Y by X** command and use them as discovery tools
- Label individual points in plots
- Produce and plot summary statistics

Contents

Look Before You Leap

When you installed JMP, a folder named **Sample Data** was also installed. In that folder is a file named Hot Dogs.jmp. Open the **Hot Dogs.jmp** file to see the data shown in Figure 3.1.

Figure 3.1 Hot Dogs Data Table

	Product Name	Type	Taste	$/oz	$/lb Protein	Calories	Sodium	Protein/Fat
1	Happy Hill Supers	Beef	Bland	0.11	14.23	186	495	1
2	Georgies Skinless Beef	Beef	Bland	0.17	21.70	181	477	2
3	Special Market's Premium Beef	Beef	Bland	0.11	14.49	176	425	1
4	Spike's Beef	Beef	Medium	0.15	20.49	149	322	1
5	Hungry Hugh's Jumbo Beef	Beef	Medium	0.10	14.47	184	482	1
6	Great Dinner Beef	Beef	Medium	0.11	15.45	190	587	1
7	RJB Kosher Beef	Beef	Medium	0.21	25.25	158	370	2
8	Wonder Kosher Skinless Beef	Beef	Medium	0.20	24.02	139	322	2
9	Happy Fats Jumbo Beef	Beef	Medium	0.14	18.86	175	479	1
10	Midwest Beef	Beef	Medium	0.14	18.86	148	375	1

The Hot Dogs.jmp table has the following information:

- The columns called Type, Calories, Sodium, and Protein/Fat (an index ratio of protein to fat) give information about nutrition. The Type column has values Meat, Poultry, and Beef.

- Cost information is in columns $/oz (dollars per ounce of hot dog) and $/lb Protein (dollars per pound of hot dog protein).

- Three categories of taste are coded Bland, Medium, and Scrumptious in the Taste column.

To get a feel for these data, use the **Distribution** command.

🖰 Select **Analyze > Distribution** to see the window in Figure 3.2.

🖰 Select all the variables *except* Product Name and click the **Y, Columns** button.

🖰 To select more than one item, highlight the first item, hold down the Shift key and press the down arrow button until all desired items are selected.

🖰 Click **OK**.

Figure 3.2 Distribution Command

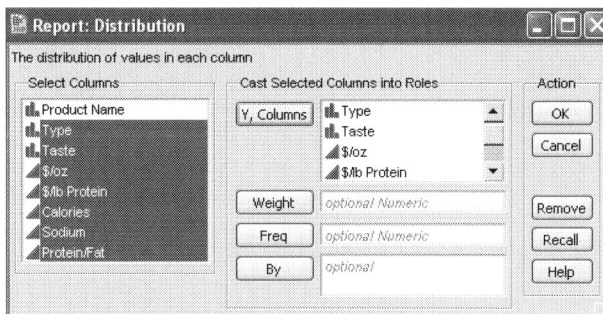

Examine the resulting report to see the distributions and levels of each variable.

Grouping Data

Of course, health is a primary concern of a school cafeteria. It is interesting to see whether the type of hot dog plays a role in healthfulness. In particular:

- Which type of hot dog has the fewest calories?
- Is the amount of sodium different in the three types of hot dogs?
- Which hot dogs have the highest protein content?
- Which hot dogs taste good and are healthy?

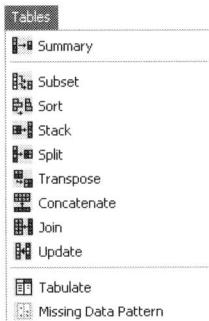

To address these issues, the data need to be grouped into hot dog type and taste preference categories with summary statistics computed for each group. The **Summary** command in the **Tables** menu groups data and computes summary statistics.

The **Summary** command creates a *summary table*. This table summarizes columns from the active data table, called its *source table*. The Hot Dogs.jmp table is the source table in this example. A summary table has a single row for each level (value) of a specified variable.

🖰 Select **Tables > Summary**.

🖰 Select **Type** and click the **Group** button to see the window as shown in Figure 3.3.

🖰 Click **OK**.

Figure 3.3 Summary Window

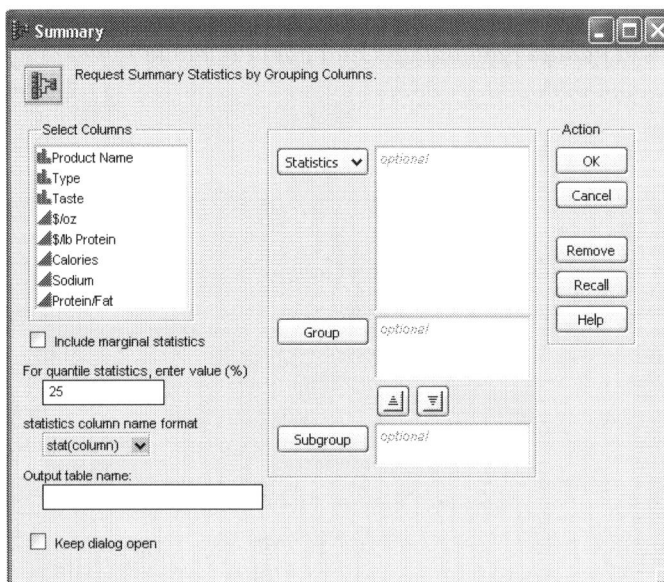

The Hot Dogs By (Type) summary table (Figure 3.4) appears in a new window. The Type column lists hot dog type and the NRows column gives the frequency of each type in the source table.

A summary table is not independent of its source table. It has these characteristics:

• When rows are highlighted in the summary table, their corresponding rows highlight in the source table.

• The summary table is not saved when closed. Select **File > Save As** to specify a name and location for the table.

Figure 3.4 Summary Table for Type of Hot Dog

	Type	N Rows
1	Beef	20
2	Meat	17
3	Poultry	17

Creating Statistics for Groups

Next, expand the summary table with columns of statistics. Summary tables have an additional command in the columns panel called **Add Statistics Column**. Use this command to add statistical summary columns to the table at any time.

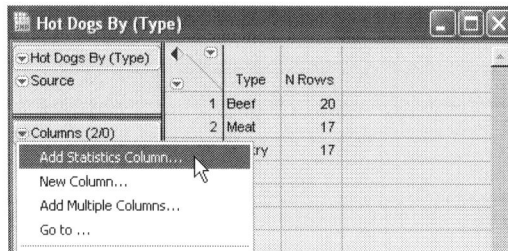

To follow along with this example, do these steps:

🖰 Click the red triangle icon and select the **Add Statistics Column** command from the columns panel on the left side of the screen.

🖰 Select Calories, Sodium, and Protein/Fat in the column selector list of the window, as shown in Figure 3.5.

🖰 Click the **Statistics** button and select **Mean**. You should now see new column names in the Statistics list.

🖰 Click **OK**.

Figure 3.5 Summary Window

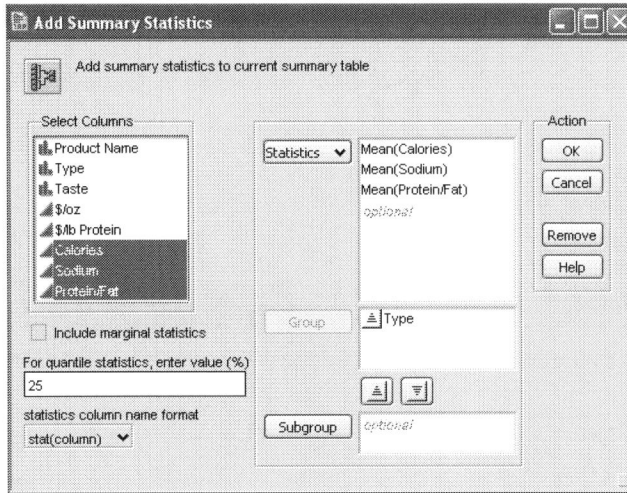

The new columns of statistics are displayed in the **Hot Dogs By (Type)** table (top table of Figure 3.6).

🖑 Repeat the previous steps to create a second summary table of **Hot Dogs by Taste** to look at health factors and hot dog tastiness. The **Hot Dogs By (Taste)** summary table shows average calories, sodium, and protein-to-fat ratio for each taste category (bottom table of Figure 3.6).

Figure 3.6 Summary Statistics for Hot Dog Groups

	Type	N Rows	Mean(Calories)	Mean(Sodium)	Mean(Protein/Fat)
1	Beef	20	156.85	401.15	1.45
2	Meat	17	158.705882	418.529412	1.41176471
3	Poultry	17	118.764706	459	3.23529412

	Taste	N Rows	Mean(Calories)	Mean(Sodium)	Mean(Protein/Fat)
1	Bland	10	172.7	466.5	1.4
2	Medium	39	139.435897	418.102564	2.20512821
3	Scrump	5	137.8	394	1.6

Charting Statistics from Grouped Data

The summary tables in Figure 3.6 show the summary statistics in tabular form, but bar charts are better for visual comparison. The **Chart** command on the **Graph** menu can also summarize data and then create charts of the summarized data.

🖑 Make sure the **Hot Dogs.jmp** table is active.

🖑 Select **Graph > Chart**.

🖑 Assign variable roles as shown in Figure 3.7.

Charts like those below should appear.

Figure 3.7 Charting Data

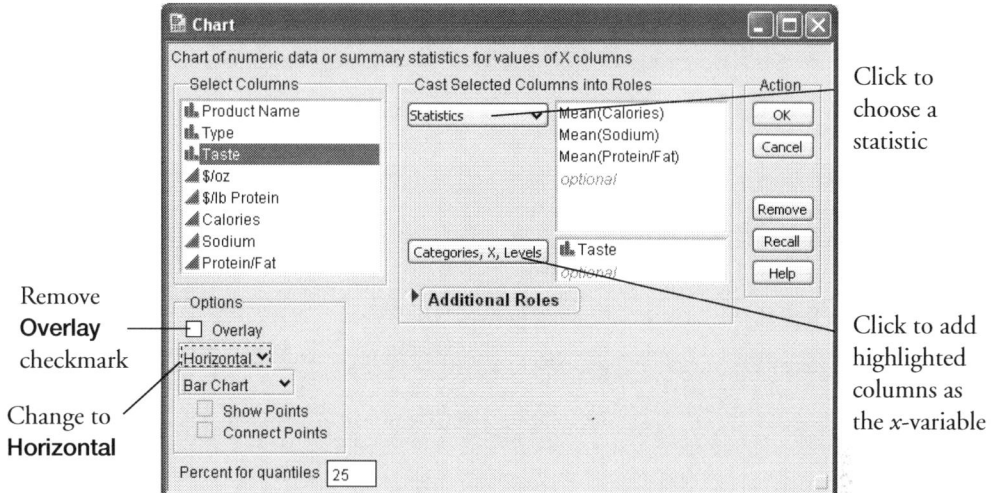

Click to choose a statistic

Remove **Overlay** checkmark

Change to **Horizontal**

Click to add highlighted columns as the *x*-variable

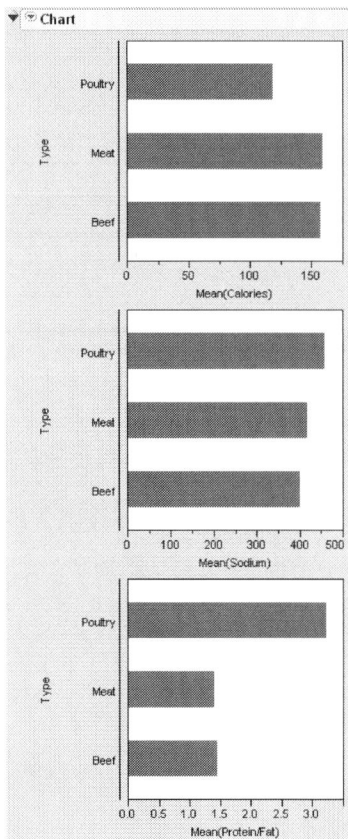

It appears that poultry hot dogs have fewer calories on average than the other two hot dog types. Also note that the poultry hot dogs have slightly more sodium. The most visible difference is that the protein-to-fat ratio appears much higher in poultry hot dogs.

🖰Repeat the above steps using **taste** to produce another set of bar charts.

🖰Select **Graph > Chart**.

🖰Select Calories, Sodium, and Protein Fat in the Select Columns list.

🖰Click **Statistics > Mean**.

🖰Select Taste from the select columns list.

🖰Click **Categories, (X), Levels**.

🖰Remove overlay checkmark.

🖰Change to horizontal.

It might not be surprising to see from the bar charts that hot dogs rated as bland tasting have (on average) more calories, more sodium, and a lower protein-to-fat ratio. It can be seen from the data table that scrumptious tasting hot dogs have the lowest average calories and sodium content. However, medium tasting hot dogs have the highest protein-to-fat ratio, and they compare well with respect to the other nutritional factors.

Charting Statistics for Two Groups

Next, it is useful to know the frequency of the three taste responses for each type of hot dog.

🖰 Make sure the Hot Dogs.jmp table is active and use **Tables > Summary** again and select both Type and Taste as grouping variables.

🖰 Click **OK**.

This produces the table shown here. There is one row for each taste response within each type of hot dog. The N Rows column lists the frequency in the source table of each type-taste combination.

	Type	Taste	N Rows
1	Beef	Bland	3
2	Beef	Medium	16
3	Beef	Scrumptious	1
4	Meat	Bland	6
5	Meat	Medium	8
6	Meat	Scrumptious	3
7	Poultry	Bland	1
8	Poultry	Medium	15
9	Poultry	Scrumptious	1

🖰 Select the **Graph > Chart** command, with both grouping variables as **Categories (X)** and the Nrows column with the *y* role.

🖰 Select **Statistics > Data**.

🖰 Click **OK**.

This produces the chart shown here. In this example, there are side-by-side charts that show the frequency for each taste within each type of hot dog.

Note: Graph > Chart can also be used to directly chart data grouped by two variables; the data doesn't have to be grouped first by **Tables > Summary**.

To label each bar with the frequency it represents:

🖰 **Label > Label by Value** is selected by default. Right-click the bars and select **Label > Show Labels**.

The chart shows that the poultry hot dogs excelled in nutrition factors and that most people find them medium-tasting. However, because the sodium content appears slightly high in some poultry brands, more investigation is needed.

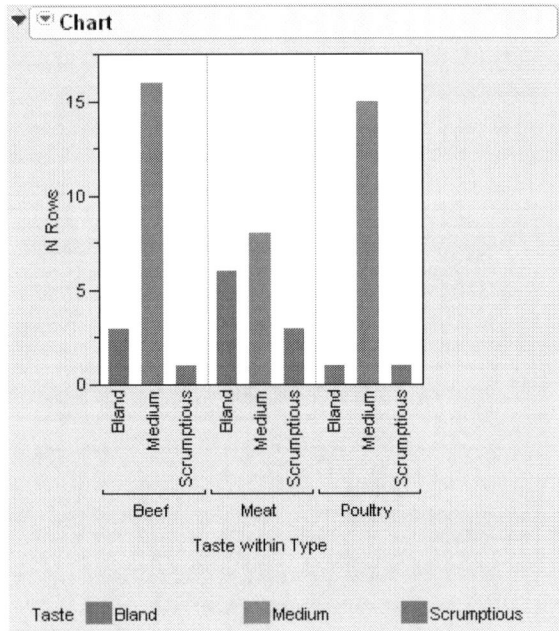

Finding a Subgroup with Multiple Characteristics

Continue the search for the ideal hot dog. Add special markers to the summary table, Hot Dogs by (Type, Taste), that identify each type of hot dog.

In the Hot Dogs by (Type, Taste) summary table:

⌐ Shift-click or click and drag over the medium and scrumptious beef rows (2 and 3) to select them.

⌐ Use the **Markers** command in the **Rows** menu to assign them the **Z** marker.

⌐ Deselect those rows.

⌐ Shift-click or drag the medium and scrumptious meat rows (5 and 6) to select them, assign them the **Y** marker, and deselect them.

⌐ Shift-click or drag the medium and scrumptious poultry rows (8 and 9), assign them the **X** marker, and deselect them.

The type-taste summary table now looks like the one shown here, and the corresponding rows in the Hot Dogs.jmp table are marked likewise.

		Type	Taste	N Rows
	1	Beef	Bland	3
z	2	Beef	Medium	16
z	3	Beef	Scrumptious	1
	4	Meat	Bland	6
Y	5	Meat	Medium	8
Y	6	Meat	Scrumptious	3
	7	Poultry	Bland	1
x	8	Poultry	Medium	15
x	9	Poultry	Scrumptious	1

Comparative Scatterplots

Now, examine the relevant variables with scatterplots to identify specific points (brands). The **Fit Y by X** command in the **Analyze** menu produces scatterplots when both the *x* and *y* are *continuous* numeric variables.

The following scatterplots graphically show the relationship of cost and the nutritional factors together.

⌐ Click the Hot Dogs.jmp source table to make it active.

⌐ Select **Analyze > Fit Y by X**.

⌐ Make your selections in the window, giving $/lb Protein the *y* role and both $/oz and Protein/Fat the *x* role.

⌐ Click **OK**.

This produces $/lb Protein by $/oz and a $/lb Protein by Protein/Fat scatterplots.

⌐ Click the red triangle icon and select **Group By**.

⌐ Choose Type as the grouping variable from the list of variables in the Grouping window.

🖑 Repeat this action for the $/lb Protein by Protein/Fat scatterplot.

🖑 For both plots, click the red triangle icon and choose **Density Ellipse > .90** to make a density ellipse visible.

🖑 Repeat to complete a similar **Fit Y by X** analysis with Calories as *y* and Sodium as *x*.

These commands produce the $/lb Protein by $/oz, the $/lb Protein by Protein/Fat, and the Calories by Sodium scatterplots shown in Figure 3.8 and Figure 3.9.

The 0.90 ellipses in the scatterplots show the shape of the bivariate response for each type of hot dog. The special markers identify the taste and type of each point.

Figure 3.8 Scatterplots Comparing Cost, Taste, and Nutritional Factors

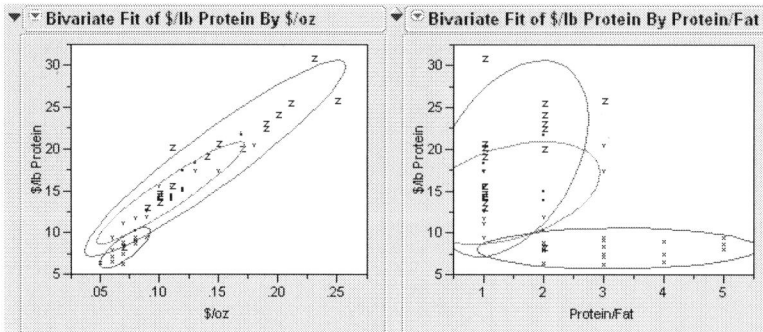

To further identify and highlight points of interest:

🖑 Select the brush tool (🖌) from the **Tools** menu.

🖑 Press the Alt key (Alt-Shift on Linux and Option on Macintosh) and drag the brush in the lower left quadrant of the Calories by Sodium scatterplot, as shown in Figure 3.9.

These points represent brands with both low sodium and low calories. The highlighted points of these healthiest brands also highlight in the other scatterplots.

Figure 3.9 Select Low Sodium and Low Calorie Brands

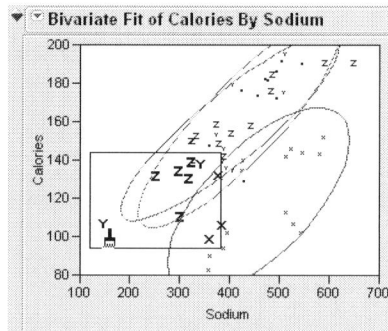

What Has Been Discovered?

The costs of meat and beef brands range from low to high. However, it is not surprising to see the tight low-cost cluster of poultry brands (**X**-marked) at the lower left of the $/lb Protein by $/oz scatterplot. The highlighted points include poultry brands, one meat brand, and one beef brand. The selected beef point (**Z**-marked) is in the upper-right corner of the plot, which places it in the most expensive category. The single meat point (**Y**-marked) is more costly than the poultry brands but less than the beef brands.

A bigger surprise appears in the $/lb Protein by Protein/Fat scatterplot. As the protein-to-fat ratio increases, the cost per pound of protein stays about the same. Further, the poultry brands not only cost the least but also contain the most protein. Most of the selected points are in the three highest protein categories.

The density ellipses on the Calories by Sodium scatterplot show clearly that the poultry brands have about the same range of sodium content as the meat and beef brands, but many poultry brands have fewer calories.

Finding the Best Points

Now there is sufficient information to identify several hot dog brands as possible cafeteria menu items.

- ⌐ Click the Hot Dogs.jmp table to make it active. Note that the icon beside the Product Name column in the Column panel means that is has been designated as a Label column.

The poultry (**X**-marked) brands are acceptably economical, and some of them have high protein content. Few meat or beef brands compared well.

- ⌐ Select the Arrow tool from the **Tools** menu.

- ⌐ Click inside the Calories by Sodium scatterplot to deselect all points.

- ⌐ Shift-click (hold down Alt-Shift and click on Linux) to highlight the two poultry brands (**X**-marked) in the data table with the least calories and lowest sodium content.

- ⌐ Shift-click to highlight the lone meat point (**Y**-marked) in the data table that has the least sodium of all brands, is low in calories, has a moderate protein count, and is average in price.

- ⌐ Select **Rows > Label/Unlabel** to display the brand names of highlighted points (Thin Jack Veal, Calorie-less Turkey, and Estate Chicken), as shown in Figure 3.10.

Figure 3.10 Labeling Ideal Points

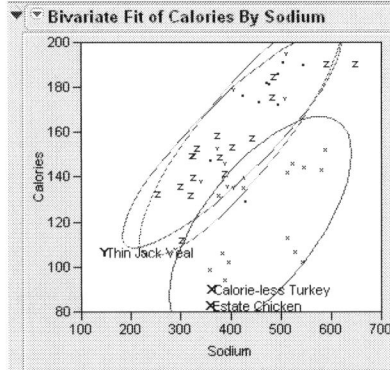

As a final step, use **Analyze > Fit Y by X** to look again at the two scatterplots that compare costs.

🖑 Select **Analyze > Fit Y by X**.

🖑 Assign $/lb Protein as **Y**.

🖑 Assign both $/oz and Protein/Fat as **X**.

🖑 Click **OK**.

The plot in Figure 3.11 shows that the Estate Chicken brand is the most economical of the three labeled brands (showing **$/oz** as continuous). The plot to the right indicates that the Calorie-less Turkey brand is in the group with the highest proportion of protein (showing **Protein/Fat** as nominal).

Figure 3.11 Winning Hot Dog Brands.

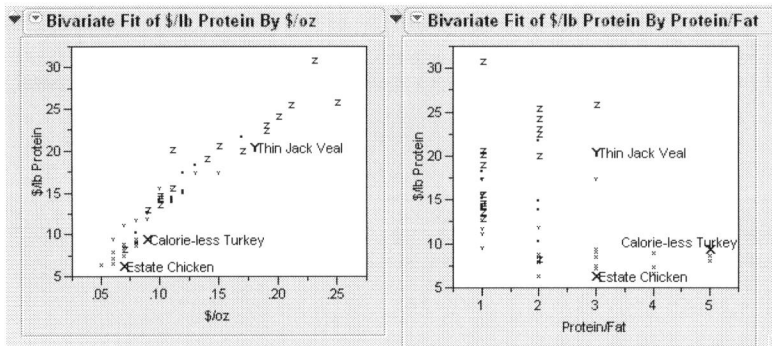

Chapter Summary

This lesson examined different hot dog brands for a cafeteria menu. A JMP table has data for 54 brands of hot dog showing type of hot dog, taste preference, nutritional factors, and cost factors.

To find the ideal hot dog, we did the following:

- Created a summary table that group the data by hot dog type and by taste preference within each hot dog type.
- Used **Graph > Chart** to chart summary statistics and identify the subset of hot dog brands that are both the most nutritious and the best tasting.
- Assigned different markers to each type of hot dog.
- Used **Analyze > Fit Y by X** to see scatterplots that compare cost factors and nutritional factors.
- Selected the points representing the lowest cost, most nutritious, and used the **Label/Unlabel** command in the **Rows** menu to identify the Calorie-less Turkey brand as a possible cafeteria hot dog.

See the *JMP User Guide* for details about the **Summary** command. For scatterplot and bar chart examples, see the *JMP Statistics and Graphics Guide*.

3 Summarizing Data

Chapter **4**

Looking at Distributions
Histograms, moments, quantiles, and proportions

The students in a local school are participating in a health study. This lesson summarizes basic information about the students for the school system's health care specialists. The data collected include age, sex, weight, and height.

To document the sample of participating students and identify any students with unusual characteristics who might need special attention, we need to view summaries of the data. This lesson produces reports with graphs and short, straightforward explanations.

Objectives

- Use the distribution analysis to explore several variables at once.
- Produce reports of moments, quantiles, frequencies, and proportions.
- Use the formula editor to compute a column's value.
- Create a subset of a data table.

Contents

Look Before You Leap

The first step in this analysis is to become familiar with the data in the Big Class.jmp file. Looking at the information in the JMP data table helps us decide which summary charts and tables to use in the health report.

🖰 Open the Big Class.jmp data table to see the data table shown in Figure 4.1.

Figure 4.1 Big Class.jmp Data Table

	name	age	sex	height	weight
1	KATIE	12	F	59	95
2	LOUISE	12	F	61	123
3	JANE	12	F	55	74
4	JACLYN	12	F	66	145
5	LILLIE	12	F	52	64
6	TIM	12	M	60	84
7	JAMES	12	M	61	128
8	ROBERT	12	M	51	79
9	BARBARA	13	F	60	112
10	ALICE	13	F	61	107
11	SUSAN	13	F	56	67
12	JOHN	13	M	65	98
13	JOE	13	M	63	105

The file contains the name, age, sex, height, and weight for each student participating in the health study. The data table is in order by age, and sex is ordered within each age group.

Even though there are only five columns of information, these variables address the following questions:

• How many boys and how many girls are there?

• How old are they?

• What is the average height and weight of the students?

• Are there any students drastically younger or older than the average age?

• Are there any students whose height or weight might signal the need for special attention?

Displaying Distributions

To summarize the data:

🖰 Select the **Distribution** command from the **Analyze** menu.

🖰 In the window that appears, select the age and sex columns as **Y, Columns**.

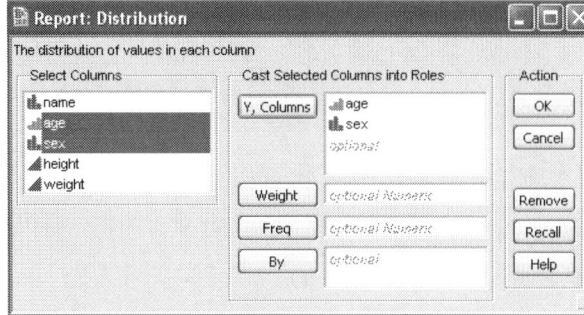

◌ Click **OK**.

The frequencies table that appears shows that the class of 40 contains 18 girls and 22 boys.

Understanding Histograms of Nominal and Ordinal Variables

After selecting **Analyze > Distribution** and completing the window, you see a window that displays histograms for analysis variables. The histogram for ordinal or nominal variables like age and sex has a bar for each level (value) of the variable.

Figure 4.2 Histogram of the age and sex Variables

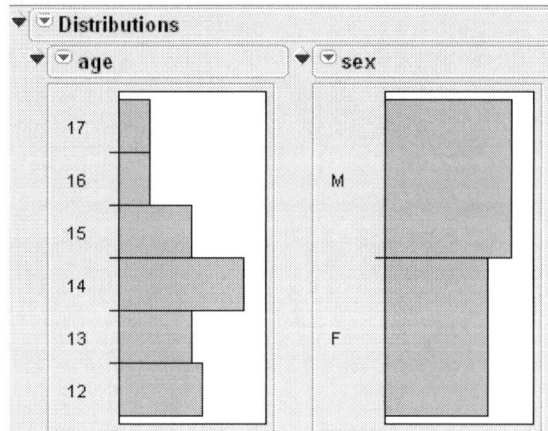

Displaying Mosaic Plots

In histograms of ordinal and nominal variables, you can display a mosaic plot by clicking the red triangle icon in the variable's title bar and selecting **Mosaic Plot**. A mosaic plot (shown in the figure to the right) visualizes the proportion of each ordinal or nominal level within the sample. It has a section for each level of the variable, where the size of the section is proportional to the corresponding group's size. Think of a mosaic plot as a bar chart with its bars stacked end to end.

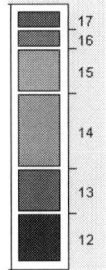

Understanding Histograms of Continuous Values

If your variables are continuous, the histogram looks slightly different than if the variables are nominal or ordinal. To see the difference, look at the distribution of the continuous variables height and weight in the sample of students.

🖰 Click the Big Class.jmp data table to make it the active window.

🖰 Again choose the **Distribution** command from the **Analyze** menu.

🖰 Designate the height and weight columns as **Y, Columns** variables.

🖰 Click **OK**.

Histograms are displayed for height and weight, as shown in Figure 4.3.

Figure 4.3 Histograms of height and weight

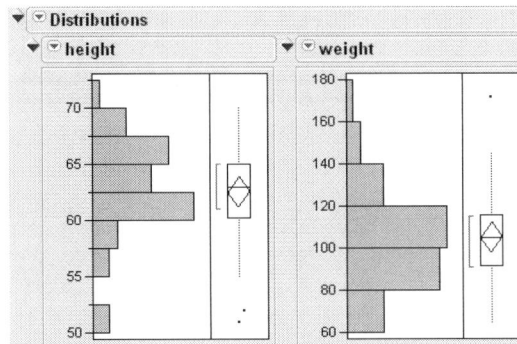

Both height and weight appear to have approximately normal (bell-shaped) distributions, but notice the extremely high weight value. It will be examined more closely later.

It is important to present data in the best possible form. Sometimes it is worthwhile to experiment with the shape of a histogram by changing the number of bars or altering their arrangement on the axis.

To adjust the histogram bars:

🖰 Select the hand from the graph cursor toolbar.

🖰 Position the hand on the bars and press the mouse button to *grab* the plot.

🖐 Move the hand to the left to increase the bar width and combine intervals (see "Graphs and Charts," p. 13). The number of bars decreases as the bar size increases.

🖐 Move the hand to the right to decrease the bar width, showing more bars.

🖐 Move the hand up or down to change the boundaries of the bins. The height of each bar adjusts according to the new number of observations within each bin.

Using Outlier Box Plots

Available by default in histograms with continuous variables, the *outlier box plot* (see Figure 4.4) is a schematic that shows the sample distribution and allows identification of points with extreme values, sometimes called *outliers*. You can display and hide an outlier box plot by clicking the red triangle icon in the variable's title bar and selecting **Outlier Box Plot**.

The ends of the box are the 25th and 75th quantiles, also called the *quartiles*. The difference between the quartiles is the *interquartile range*. The line across the middle of the box identifies the *median* sample value.

The lines extending from each end of the box are sometimes called *whiskers*. The whiskers extend from the ends of the box to the outermost data points that fall within the distance computed as `quartile ±` `1.5*(interquartile range)`. Points beyond the whiskers indicate extreme values that are possible outliers. To label a point, click the point to highlight it, and then select **Rows > Label/Unlabel**.

The red bracket along the edge of the box identifies the *shortest half*, which is the most dense 50% of the observations.

Figure 4.4 Outlier Box Plot

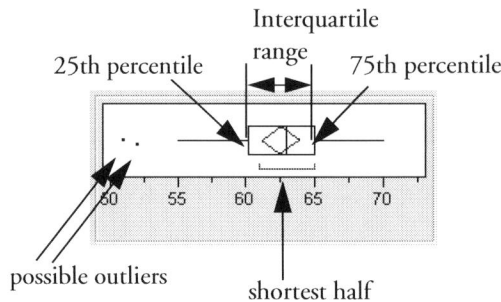

Using Quantile Box Plots

In histograms whose variables are continuous, you display a quantile box plot by clicking the triangle icon in the variable's title bar and selecting **Quantile Box Plot**.

A quantile box plot shows the location of preselected percentiles, sometimes called *quantiles*, on the response axis. The median shows as a line in the body of the box. The ends of the box locate the 25th and 75th quantiles. The number of other quantile lines depends on the available space. The accompanying text report lists the data values for each of the standard quantiles. The box also contains a *means diamond*. The two diamond points within the box identify the 95% confidence interval of the mean. The line that passes through the two diamond points spanning the box identifies the sample mean.

Looking at the quantile box plot and means rectangle together helps see if data are distributed normally, as shown in Figure 4.5. If data are distributed normally (bell shaped), then the 50th quantile and the mean are the same and other quantiles show symmetrically above and below them.

Figure 4.5 Quantile Box Plot and Quantiles Table

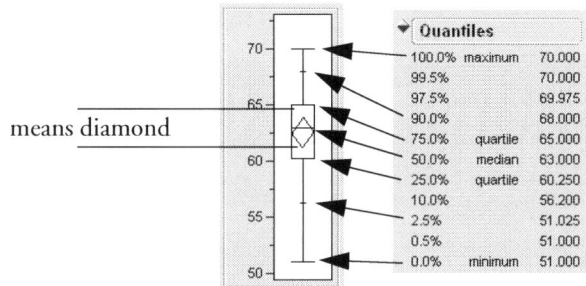

Learning About Report Tables

JMP produces tables of statistical summaries along with graphs. The tables that JMP produces depend on whether a variable is continuous, ordinal, or nominal. Click the red triangle icon and select **Display Options** to reveal tables that are not available by default.

The diamond-shaped disclosure button (◆ ◆ on Windows and Linux and ▶ ▼ on the Macintosh) at the top-left of each report opens and closes it.

Reports for Continuous Variables

The figure to the right is the table JMP gives along with histograms for the continuous variable height.

- The Quantiles table displays the maximum value, minimum value, and other values for selected quantiles.

- The Moments table displays the mean, standard deviation, and other summary statistics.

Note: You can also obtain more moments by clicking the red triangle icon on the variable's title bar and choosing **Display Options > More Moments**.

Frequency Table for Ordinal or Nominal Variables

The report for nominal and ordinal variables has a different table from those produced for continuous variables. The table JMP gives along with histograms for the nominal or ordinal (categorical) variables sex and age has frequency tables that show these items:

Frequencies				
Level	Count	Prob	StdErr Prob	Cum Prob
12	8	0.20000	0.06325	0.20000
13	7	0.17500	0.06008	0.37500
14	12	0.30000	0.07246	0.67500
15	7	0.17500	0.06008	0.85000
16	3	0.07500	0.04165	0.92500
17	3	0.07500	0.04165	1.00000
Total	40	1.00000	0.00000	1.00000
N Missing	0			
6 Levels				

- **Level**   lists each value of the response variable.
- **Count**   lists the number of rows found for each level of a response variable.
- **Prob**   lists the probability of occurrence for each level of a response variable. The probability is computed as the count divided by the total frequency of the variable, shown at the bottom of the table.

The following two statistics are not displayed by default. Right-click (hold the CONTROL key and click on the Macintosh) the frequency table and select the **Columns** menu to reveal them.

- **StdErr Prob**   lists the standard error of the probabilities.
- **Cum Prob**   contains the cumulative sum of the column of probabilities.

Adding a Computed Column

Be on the alert for any unusual subjects, such as students who have extreme height or weight values. A good indicator of extreme values is the ratio of weight to height. To examine the ratio of weight to height, create a new column called ratio, computed as weight divided by height. To do this:

🖑 Click the Big Class.jmp data table to make it the active window.

🖑 Select **Cols > New Column**.

To create the new column of weight-to-height ratios, complete the New Column window as in Figure 4.6.

🖑 Type the new name, ratio, in the **Column Name** area.

The default data type is **Numeric** and is correct as is.

The modeling type is **Continuous** and is correct as is.

🖑 Click the drop-down menu beside **Format** and set the format for ratio in the data grid to **Fixed Dec** with two decimal places.

🖑 Click the **Column Properties** button and select **Formula**, as shown in Figure 4.6, to compute values for the new column.

Figure 4.6 New Column Window

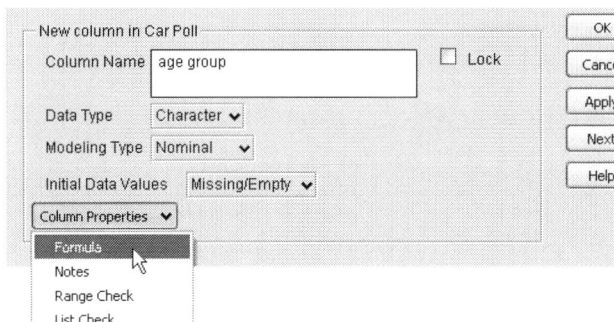

Construct the formula that calculates values for the ratio column as follows:

🖑 Highlight the empty term in the formula and select **weight** from the list of column names in the upper-left corner of the formula editor.

🖑 Press the divide (÷) key on the formula editor keypad.

🖑 With the empty denominator term highlighted, select **height** from the list of column names.

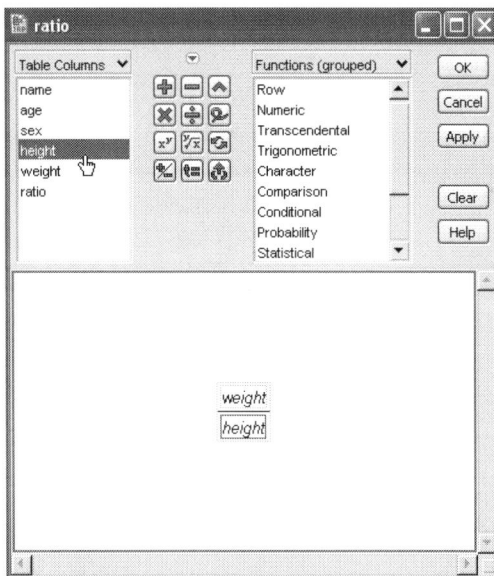

🖑 When the formula is complete, click **Apply** or **OK** on the formula editor, or just close its window.

The new column called ratio is now in the Big Class data table as shown here. Its values are the computed weight-to-height ratio for each student.

Now look at the distribution of the ratio variable.

🖰 Select **Analyze > Distribution** and assign the new column (ratio) to the **Y, Columns** role.

🖰 Click **OK**.

One way to identify subjects that have extreme values is to highlight histogram bars for the highest and lowest values. To highlight more than one bar, press the Shift key and click the desired bars.

🖰 Select the bars for the two lowest and one highest bar (Figure 4.7).

Figure 4.7 Histogram of Ratio with Bars Highlighted at Extreme Values

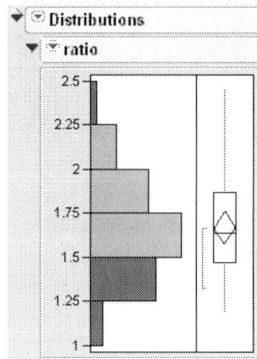

The highlighted bars in the histogram represent a ratio either greater than or equal to 2.25 or less than 1.5. The corresponding points automatically highlight in the data table and in all other reports generated from the Big Class data table.

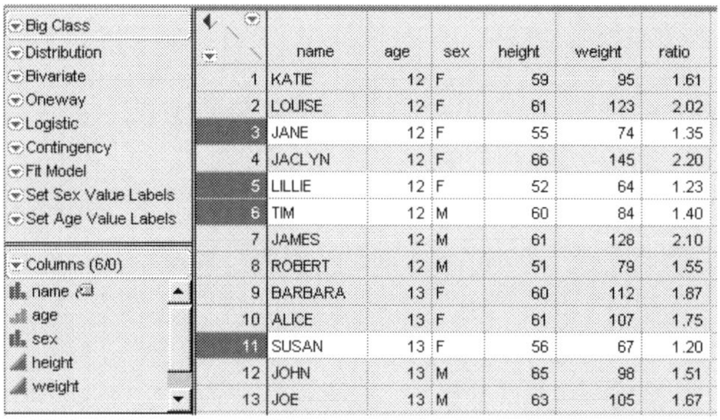

Creating Subsets

Looking in the Big Class data table allows examination of the selected rows, but scrolling through a large data grid can be tedious. For the final report to the health researchers, include a separate list containing only the highlighted students—those with extreme values. To do this, use **Tables** menu commands to create new data tables or modify existing tables.

🖑 Select **Tables > Subset** or click the Big Class red triangle icon and select **Tables > Subset**.

🖑 Click **OK** to accept the default choices presented in the window.

This creates a new data table that has only the selected rows and columns from the active data table.

The new data table, shown in Figure 4.8, contains only the students that have extreme weight-to-height ratios. By default, the table is named Subset of Big Class. Change the name by clicking the existing name (Subset of Big Class) in the panel located on the top left side of the window. The table can be saved, exported for use in another application, or printed.

Figure 4.8 Data Table Containing a Selected Subset

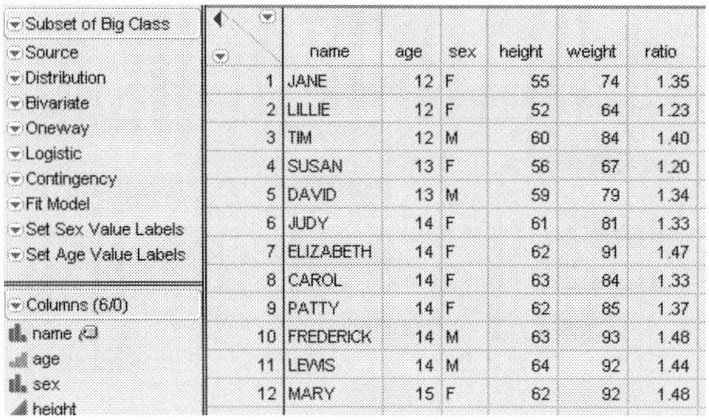

Chapter Summary

In this chapter, the demographic and vital data of students participating in a health study were summarized. The profile was completed using the **Distribution** command and the data management features of the JMP data table.

The **Distribution** command displayed histograms and box plots or stacked (mosaic) bar charts for each variable assigned the role of response variable (*y*). Using display and text report options to look more closely at the data, the following actions were completed:

- Adjusted the number of bars and the scale of the histograms
- Produced supporting statistical reports showing moments and quantiles of numeric variables and frequencies and proportions of nominal and ordinal variables
- Created a new column in the data table computed as a function of existing columns
- Highlighted histogram bars to identify a subset of rows in the data table
- Created a new data table from a subset of highlighted rows

Graphs and text reports can be printed directly from JMP. Graphs and reports can be copied to a JMP journal or into other applications to complete a report for the school system health care specialists.

See the chapter "Univariate Analysis" in the *JMP Statistics and Graphics Guide* for more information about distributions.

Chapter **5**

Comparing Group Means
Testing Differences

The company has decided to replace all computer keyboards with the brand that produces the fastest accurate typing. The employees participated in a study to help decide what type of keyboards to buy. The company selected three different brands of keyboards to test. These keyboards were randomly assigned to three groups of employees with comparable typing skills. The employees completed typing tests and recorded their words-per-minute scores.

This lesson finds out if the typing scores are significantly better on any one brand of keyboard than on the others.

Objectives

- Use the **Fit Y by X** command to produce plots and analyses appropriate for a one-way analysis of variance.
- Use interactive tools to examine differences among groups.
- Produce text reports to display differences among groups.

Contents

Look Before You Leap

The first step is to become familiar with the data. The typing test scores are in a JMP file so that they can be reviewed and the type of analysis determined.

🖰 When you installed JMP, a folder named Sample Data was also installed. In that folder is a file named Typing Data.jmp. Open the file Typing Data.jmp.

The Typing Data table appears in the form of a data grid, as shown here.

			brand	speed
▾ Typing Data				
Notes Three brands of typ		1	REGAL	70
▾ Oneway		2	SPEEDYTYPE	87
▾ Columns (2/0)		3	SPEEDYTYPE	79
▮ brand		4	REGAL	73
◢ speed		5	SPEEDYTYPE	77
▾ Rows		6	REGAL	72
All rows	17	7	WORD-O-MATIC	62
Selected	0	8	REGAL	71
Excluded	0	9	WORD-O-MATIC	77
Hidden	0	10	SPEEDYTYPE	80
Labelled	0	11	REGAL	72

The data table has columns named brand and speed. The modeling type for each column shows to the left of each column name in the columns panel. The character variable brand has nominal (▮) values and the numeric variable speed has continuous (◢) values.

There are 17 rows that represent typing scores for 17 employees. However, the number of participants in the groups differs because some of the scheduled participants did not show up for the study. Perhaps other statistics for the groups differ also. In particular:

- Is the mean (average) typing speed the same for each brand?
- Do any one of the three brands of keyboard stand out from the others?
- Does it make a difference as to which brand the employees use?

Graphical Display of Grouped Data

Comparing the mean typing scores of each keyboard brand involves analyzing two variables, so use the **Fit Y by X** command from the **Analyze** menu.

Selecting **Fit Y by X** allows you to perform:

- Categorical analysis when both *x* and *y* have nominal or ordinal values
- Analysis of variance when *x* is nominal or ordinal and *y* has continuous values, as in the example shown here
- Logistic regression when *x* is continuous and *y* has nominal or ordinal values
- Regression analysis when both *x* and *y* have continuous values.

Choose Variable Roles

To discover if typing speed is related to (dependent on) a brand of keyboard, follow these steps:

🖑 Choose **Analyze > Fit Y by X**.

🖑 Select brand as **X, Factor** and speed as **Y, Response**. (See Figure 5.1.)

🖑 Click **OK**.

The plot shown in Figure 5.2 appears.

Figure 5.1 The Fit Y by X Launch Window

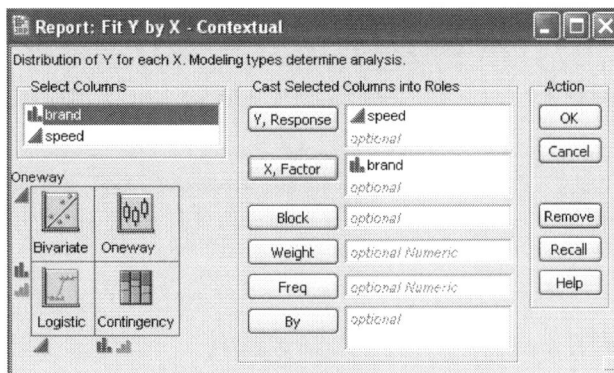

Selecting **Fit Y by X** and completing the window produces a statistical analysis appropriate for the variable roles (*x* and *y*) and the modeling type (continuous and nominal or ordinal) of each variable.

• **Y, Response**   identifies a response (dependent) variable.

• **X, Factor**   identifies a classification (independent) variable.

The next step is to choose an analysis that investigates if there is a statistical difference between the group mean values.

Show Points

Each of the typing test scores is plotted for each brand of keyboard. Note that the distance between tick marks on the **brand** axis is proportional to the sample size of each group. The mean typing score for the total sample is shown as a horizontal line across the plot.

Figure 5.2 Oneway Analysis for speed by brand

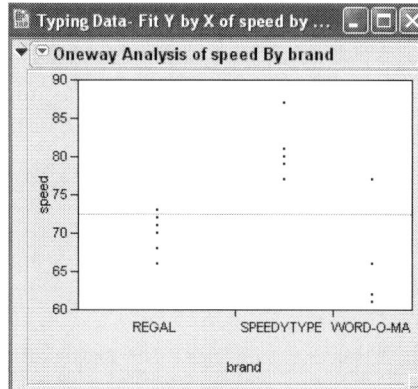

It is easy to see at a glance that most participants who used the SPEEDYTYPE machines typed faster than the others.

Fit Means Option

Now look at more graphical information about the distribution of typing scores.

🖑 Click the red triangle icon on the title bar and select **Means/Anova**.

This produces the appropriate analysis of variance reports. It automatically turns on mean diamonds, which draws a 95% means diamond for each group, as shown in Figure 5.3. You can also turn the **Mean Diamonds** option on and off from the **Display Options** submenu. Click the red triangle icon on the title bar and select the **Display Options** submenu. Mean Diamonds has a check mark beside it if it is turned on.

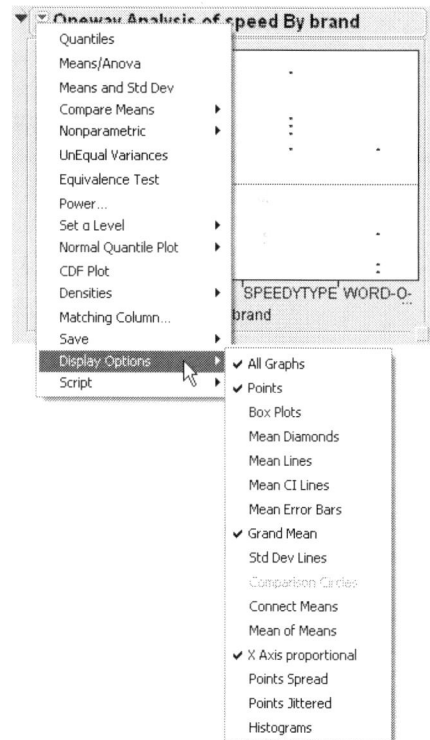

Figure 5.3 Example of Means Diamonds

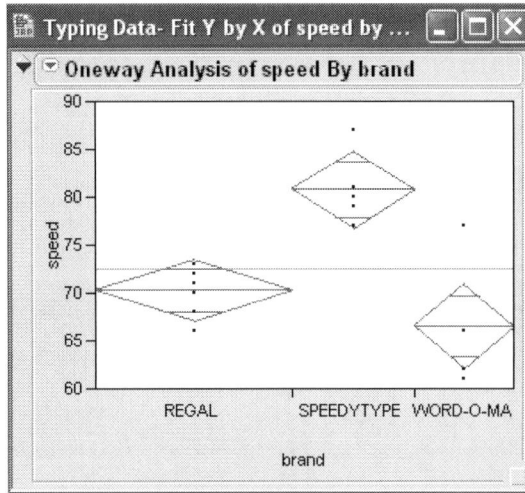

The illustration in Figure 5.4 illustrates the means diamond.

- The means diamond has a line drawn at the mean (average) value of words-per-minute for each brand of keyboard.
- The upper and lower points of the means diamond span a 95% confidence interval computed from the sample values for each machine.
- The width of each diamond spans the distance on the horizontal axis proportional to the group size.

Overlap lines within each diamond are drawn at $(\sqrt{2}/2) \cdot (CI/2)$. For groups with equal sample sizes, the marks that appear not to overlap indicate that two group means *could* be significant at the 95% confidence interval.

Figure 5.4 Means Diamond with **X Axis Proportional** Option Turned On (Left) and Off (Right)

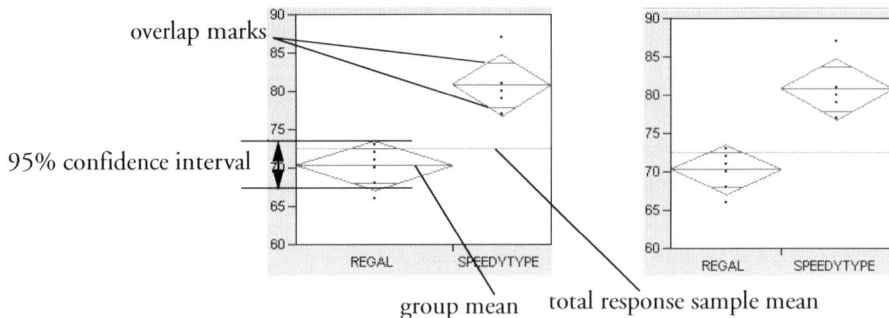

The mean scores of the **REGAL** and **WORD-O-MATIC** keyboards appear to be nearly the same, but note that the **SPEEDYTYPE** mean is much higher (Figure 5.3).

Fit Quantiles

The next logical step is to check the distribution of points within each group. This gives a better idea of the spread of the values and shows the distance of extreme values from the center of the data.

🔍 Click the red triangle icon and select **Quantiles**.

When you select the **Quantiles** command, JMP automatically overlays a quantile box plot on each group of typing scores, as shown in Figure 5.5. JMP also displays the report in Figure 5.5, which lists the standard percentiles for each keyboard. The median (50th percentile) is the typing speed that divides the sample in half. This means that 50% of the employees had speeds greater than the median, and the other half had lower speeds.

Figure 5.5 Fit Quantiles Option

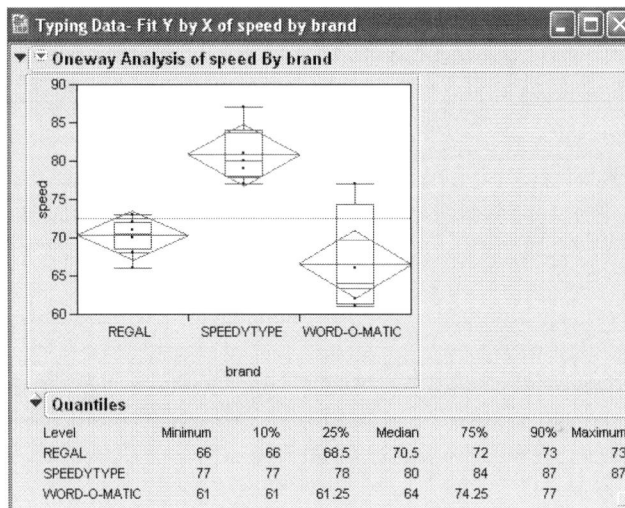

Level	Minimum	10%	25%	Median	75%	90%	Maximum
REGAL	66	66	68.5	70.5	72	73	73
SPEEDYTYPE	77	77	78	80	84	87	87
WORD-O-MATIC	61	61	61.25	64	74.25	77	

Figure 5.6 illustrates the quantile box plot. The median, or 50th quantile, shows as a line in the body of the box. The top and bottom of the box represent the 75th and 25th quantiles, also called the upper and lower quantiles. The box encompasses the interquantile range of the sample data. The 10th and 90th quantiles show as lines above and below each box.

Looking at the quantile box plot and the means diamond together helps show if data are distributed normally within a group. If data are normally distributed (bell shaped), the 50th percentile and the mean are the same and the other quantiles are arranged symmetrically above and below the median.

Figure 5.6 Quantiles Box Plot

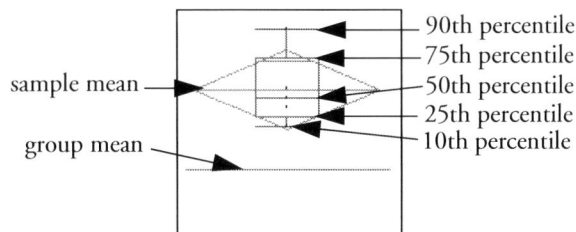

The quantile box plots (Figure 5.5) show a difference in variation of scores across the three groups. The scores in the REGAL group cluster tightly around the mean score but the WORD-O-MATIC scores show much more variation. However, even with this variation among the groups, the SPEEDYTYPE brand still appears to promote the best performance.

Comparison Circles

To complete the typing data inspection:

🖑 Click the red triangle icon and choose **Compare Means > All Pairs, Tukey HSD**.

This option produces statistical reports (discussed later) and automatically draws a set of *comparison circles* to the right of the plot that provides a graphical test of whether the mean typing scores are statistically different. Comparison circles for the three word-processor groups are shown in Figure 5.7.

The center of each circle is aligned with the mean of the group it represents. If you select Student's *t*-test instead, the diameter of each circle spans the 95% confidence interval for each group. Whenever two circles intersect, the confidence intervals of the two means overlap, suggesting that the means *might* not be significantly different. Whenever two circles do not intersect, the group means they represent are significantly different.

🖑 Click the SPEEDYTYPE comparison circle.

This graphically illustrates that the SPEEDYTYPE machine is statistically better than the other machines. The comparison circles highlight to show the statistical magnitude of the difference between typing scores. Circles for groups that are statistically the same have the same color.

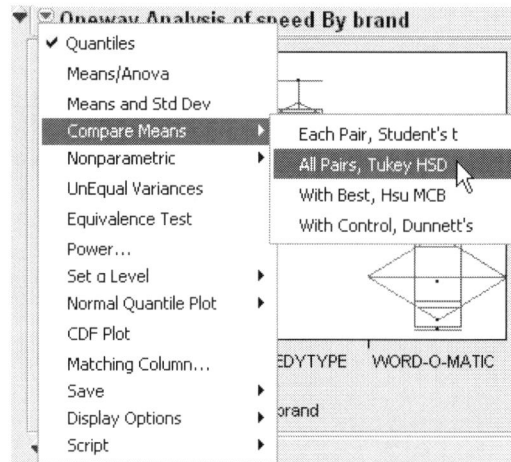

Figure 5.7 Comparison Circles

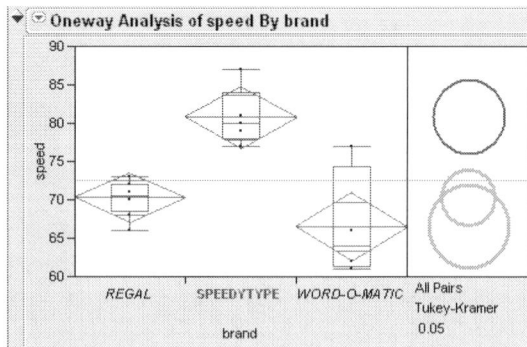

The comparison circle for the SPEEDYTYPE brand does not intersect with either of the other two. The REGAL and WORD-O-MATIC brands are statistically slower than SPEEDYTYPE but do not appear different from each other. A later section, "Mean Estimates and Statistical Comparisons," p. 66, discusses the multiple comparison tests the comparison circles represent.

Quantify Results

Now, examine the report beneath the plot that consists of several tables. The Summary of Fit table, shown in Figure 5.8, summarizes the typing data distribution with these statistics:

- **Rsquare (R^2)** quantifies the proportion of total variation in the typing scores resulting from different keyboards rather than from different people.
- **Rsquare Adj** adjusts R^2 to make it more comparable over models with different numbers of parameters.
- **Root Mean Square Error (RMSE)** is a measure of the variation in the typing scores that can be attributed to different people rather than to different machines.
- **Mean of Response** is the mean (average) of all the typing scores.
- **Observations** is the total number of scores recorded.

Figure 5.8 Summary of Fit Report for speed By brand

Oneway Analysis of speed By brand	
Oneway Anova	
Summary of Fit	
Rsquare	0.67446
Adj Rsquare	0.627954
Root Mean Square Error	4.27033
Mean of Response	72.47059
Observations (or Sum Wgts)	17

Analysis of Variance

When you select the **Means/Anova** command from the red triangle icon in the title bar, JMP gives you a standard analysis of variance table. If there are only two group levels, the report also includes a *t*-test table.

Note that the value of the *F*-probability (Prob>F) for the Analysis of Variance is 0.0004. This implies that differences as great as seen in this typing trial are expected only four times in 10,000 similar trials if the keyboards did not really promote different typing performances.

The Analysis of Variance table has the following information:

- **Source** lists the sources of variation: brand, Error, and C. Total.
- **DF** is the degrees of freedom associated with the three sources of variation.
- **Sum of Squares** (SS for short) identifies the sources of variation in the typing scores.
- **C. Total** is the corrected total SS. It divides (partitions) into the SS attributable to brand and the SS for Error. The brand SS is the variation in the typing scores explained by the analysis of variance model, that hypothesizes the keyboards are different. The Error SS is the remaining or unexplained variation.

- **Mean Square**   is the sum of squares divided by its associated degrees of freedom.
- **F Ratio**   is the model mean square divided by the error mean square.
- **Prob > F**   is the probability of obtaining a greater F-value if the mean typing scores for the keyboards differed only because different people were typing on them rather than because the keyboards promoted different scores in any way.

▼ **Oneway Anova**
▶ **Summary of Fit**
▼ **Analysis of Variance**

Source	DF	Sum of Squares	Mean Square	F Ratio	Prob > F
brand	2	528.93529	264.468	14.5027	0.0004*
Error	14	255.30000	18.236		
C. Total	16	784.23529			

Mean Estimates and Statistical Comparisons

To see the list of means for each group, look at the Means for Oneway Anova table. This table summarizes the scores for each brand and reveals what level of performance to expect.

The Means for Oneway Anova table shows the following information:

- **Level**   lists the name of each group.
- **Number**   is the number of scores in each group.
- **Mean**   is the mean of each group.
- **Std Error**   is the standard error of each group mean.
- **Lower 95%**   is the lower 95% confidence interval for the group means.
- **Upper 95%**   is the upper 95% confidence interval for the group means.

▼ **Oneway Anova**
▶ **Summary of Fit**
▶ **Analysis of Variance**
▼ **Means for Oneway Anova**

Level	Number	Mean	Std Error	Lower 95%	Upper 95%
REGAL	8	70.2500	1.5098	67.012	73.488
SPEEDYTYPE	5	80.8000	1.9097	76.704	84.896
WORD-O-MATIC	4	66.5000	2.1352	61.921	71.079

Std Error uses a pooled estimate of error variance

When you select the **Compare Means** command from the red triangle icon in the title bar, JMP gives several multiple comparison options to statistically compare pairs of groups. This example uses the **All Pairs, Tukey HSD** option, which performs a statistical means comparison for the three pairs of means using the *Tukey-Kramer HSD* (honestly significant difference) test (Tukey 1953, Kramer 1956). This means comparison method compares the actual difference between group means with the difference that would be significantly different. The difference needed for statistical significance is called the LSD (least significant difference).

The graphical results show as the comparison circles previously seen in Figure 5.7. The circles' centers represent the actual difference in the group means. The corresponding report is the Means Comparisons table (Figure 5.9), which shows the actual absolute difference between each mean and the LSD. The top half of the report gives information based on a Student's *t* comparison of each pair. The bottom half shows the results of the Tukey-Kramer multiple comparison tests. Pairs with a positive value are significantly different. The Means Comparison table confirms the visual results in Figure 5.7.

Figure 5.9 Means Comparisons Table for Tukey-Kramer HSD

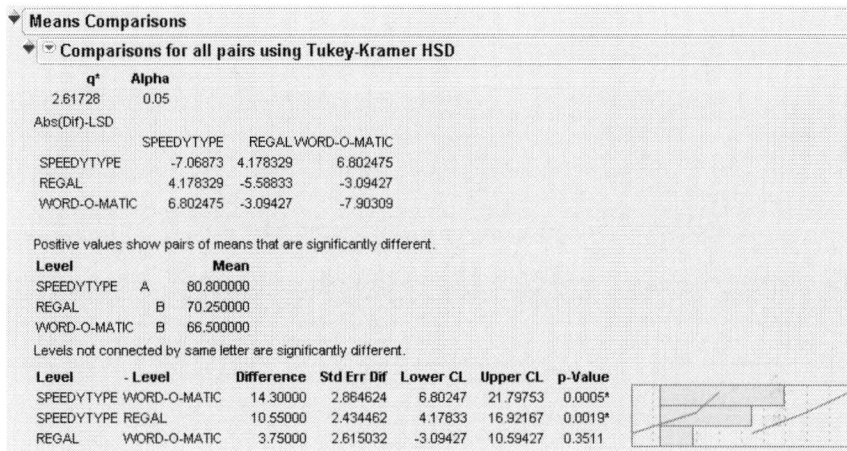

Chapter Summary

In this chapter, the difference in mean typing scores for three brands of keyboard was summarized using the **Fit Y by X** command in the **Analyze** menu. This command was also used to:

- Plot the typing scores for the three brands of keyboard.
- Overlay a means diamond on each group of typing scores to compare the means of each group.
- Overlay a quantile box plot on each group of typing scores to compare the shape of the distribution of scores in each group.
- Produce comparison circles to visualize the difference in mean typing scores.
- Compute and display a one-way analysis of variance table, which confirmed that at least one pair of means is statistically different.
- Display a table of the group means and standard errors.
- Display a table showing the multiple comparison statistical test results for group means.

Using the selection tool (⊕) from the **Tools** menu, the graphs or tables can be copied and prepared in a report for presentation. The analysis concludes that, in this typing trial, the SPEEDYTYPE keyboard produced significantly higher scores than either of the other two brands.

See the chapter "Oneway Layout" of the *JMP Statistics and Graphics Guide* for a complete discussion of one-way analysis of variance.

Chapter **6**

Analyzing Categorical Data
Comparing Proportions

Survey data are frequently categorical data rather than measurement data. Analysis of categorical data begins by simply counting the number of responses in categories and subcategories. Counting is easy, but interpreting the relationship between categories based on counts is more complex. It requires computing probabilities and evaluating the likelihood of these probabilities compared to expectations.

For example, an American automobile manufacturer—feeling the pinch of competition from foreign auto sales—needs a market analysis before proceeding with a multimillion-dollar advertising campaign. A random sample of people is surveyed. The auto manufacturer wants to know each participant's age, sex, marital status, and auto information. The auto information consists of the manufacturing country, the car's size, and the car's type, and whether it is a family, work, or sporty car. This information might provide the advertising experts with direction for the upcoming advertising campaign.

Who buys what?

Objectives

- Use the **Fit Y by X** command to compare two variables consisting of categorical data.

- Use the formula editor to re-code a categorical variable as a numeric variable.

- Produce and examine graphs and statistics appropriate for the comparison of proportions such as Chi-squared tests and mosaic plots.

Contents

Look Before You Leap

The first step is to become familiar with the data. Begin by reviewing the data to determine the best way to proceed with the market analysis.

Open a Data Table

🖰 When you installed JMP, a folder named Sample Data was also installed. In that folder is a file named Car Poll.jmp. Open Car Poll.jmp, as shown here.

The Car Poll data were collected from a random sample of people in a specific geographic area. The columns panel shows that Age is a numeric variable and is assigned the continuous (◢) modeling type. The

		sex	marital status	age	country	size	type
	1	Male	Married	34	American	Large	Family
	2	Male	Single	36	Japanese	Small	Sporty
	3	Male	Married	23	Japanese	Small	Family
	4	Male	Single	29	American	Large	Family
	5	Male	Married	39	American	Medium	Family
	6	Male	Single	34	Japanese	Medium	Family
	7	Female	Married	42	American	Large	Family
	8	Female	Married	40	European	Medium	Family
	9	Male	Married	28	American	Medium	Sporty
	10	Female	Married	26	American	Medium	Family
	11	Female	Married	26	European	Small	Sporty
	12	Male	Single	26	European	Medium	Sporty

other five columns are character variables with nominal (▮▮) modeling types.

Address the Research Question

The basic research question asks, "Is the response probability for country of manufacture, size of car, or type of car a function of the age, sex, or marital status of the owner?"

Look at the data table to see what specific relationships lend insight into this question. You are interested in the relationships between the following automobile characteristics and demographics:

- manufacturing country by age
- manufacturing country by sex
- manufacturing country by marital status
- size of car by age
- size of car by sex
- size of car by marital status
- type of car by age
- type of car by sex
- type of car by marital status

Modify the Data Table

You can sometimes obtain better summary information from age groups rather than specific ages. In fact, dividing people into two age groups is often the basis for a valuable broad analysis. So, let's find the median age—the age that divides the sample into two equal age groups.

The distribution of a variable and its corresponding quantiles display a good way to form sample groups. Use the distribution of the **age** column to find a reasonable value of age that divides the sample into two groups.

◌ Select **Analyze > Distribution**.

◌ When the Distribution window appears, select **age** as the analysis column (**Y, Column**) and click **OK**.

JMP displays a histogram with an accompanying outlier box plot, Quantiles table, and Moments table. The Quantiles table, shown here, identifies 30 as the median age.

Quantiles		
100.0%	maximum	60.000
99.5%		56.880
97.5%		44.400
90.0%		38.000
75.0%	quartile	35.000
50.0%	median	30.000
25.0%	quartile	26.000
10.0%		24.000
2.5%		22.000
0.5%		19.040
0.0%	minimum	18.000

The next step is to create a new column whose values identify whether a subject's age is greater than 30, or is less than or equal to 30.

◌ Select **Cols > New Column** to display the New Column window, which is used to define column characteristics.

Data Type, **Modeling Type**, and **Format** options define the new column's characteristics. Enter characteristics for the new column as follows:

◌ Type the new name (call it **age group**) in the **Column Name** text box.

◌ Because the new column has grouping values instead of measurements, select **Character** from the box beside **Data Type**.

Figure 6.1 New Column Window

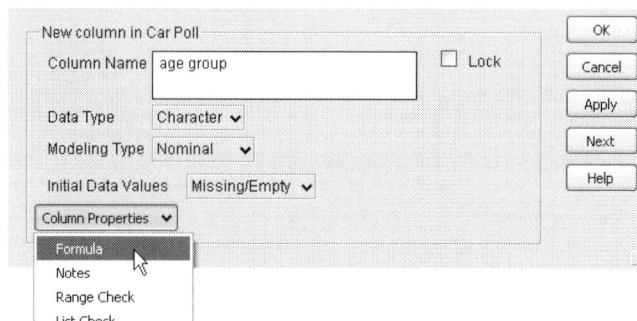

◌ Click the **Column Properties** button and select **Formula**. (See Figure 6.1.)

You are presented with the formula editor window shown in Figure 6.2.

Figure 6.2 Formula Editor Window

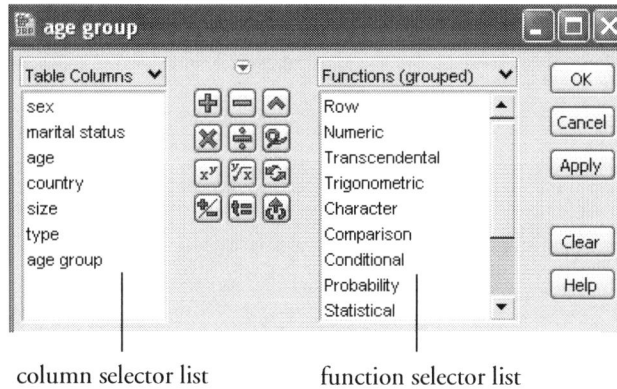

column selector list   function selector list

Suppose 0 represents ages greater than the median (30) and 1 represents the ages less than or equal to the median. To create a formula that divides the sample into two groups, follow these steps:

🖱 Click **Conditional** in the function selector list and select the **If** function.

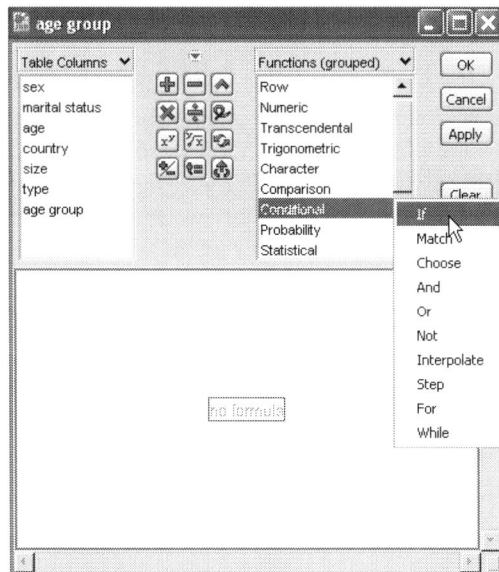

🖱 Highlight the expression term, denoted **expr**.

🖱 Choose **a<=b** from the **Comparison** functions.

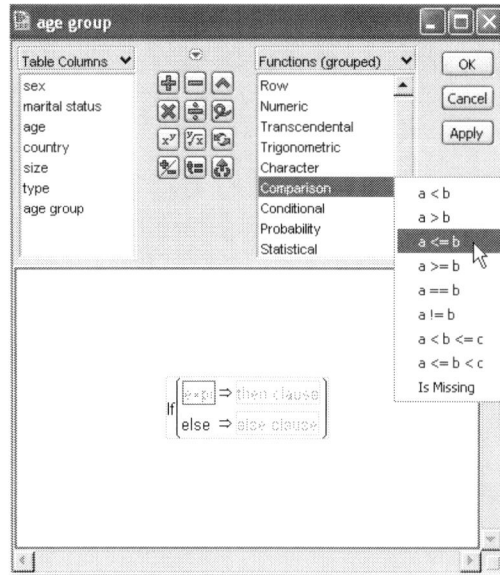

🖱 Highlight the left side of the comparison clause and click **age** in the **Table Columns** list.

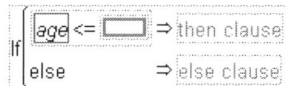

🖱 Double-click the right side of the comparison clause to obtain a text entry box.

🖱 Enter **30** for the numeric comparison.

🖱 Double-click the term denoted **then clause**.

🖱 Enter **"1"** (in double quotes because this column is a character variable).

🖱 Double-click the term denoted **else clause**.

🖱 Enter **"0"** (with double quotes).

The complete equation should look like the one shown here.

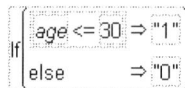

🖱 Click **Apply**, **OK**, or the formula editor's close box to fill the new column with calculated values.

Tip: Instead of using the buttons in the formula editor, you can double-click the outermost nesting box to create a single text entry box and enter **if(age<=30, "1", "0")**. Then, press Enter (or Return) or click outside the text box, and the formula appears in formatted form.

Contingency Table Reports

The nominal age grouping variable shows the relationship of age to the other nominal variables using contingency tables. To look at combinations of two variables:

🖱 Choose **Analyze > Fit Y by X**.

JMP does the statistical analysis appropriate for a variable's modeling types and role assignments.

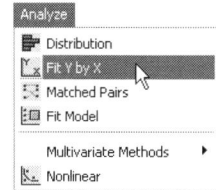

Cast Variables Into Roles

Assign analysis roles to variables by choosing an analysis from the **Analyze** menu and making selections in the window that appears. In this investigation, the country, size, and type columns are dependent response (*y*) variables; sex, marital status, and age group are independent (*x*) variables. This example shows how to complete the **Fit Y by X** window (Figure 6.3).

🖱 Select the three *y* variables (country, size, and type).

🖱 Click the **Y, Response** button.

🖱 Assign the *x* variables (sex, marital status, and age group) by selecting them and clicking the **X, Factor** button.

🖱 Click **OK** when finished.

Figure 6.3 The Fit Y by X Window

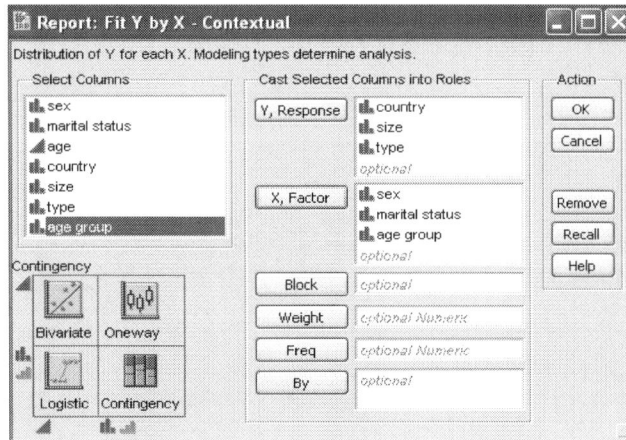

Contingency Table Mosaic Plots

If both *x* and *y* have either nominal or ordinal values, JMP displays a mosaic plot with accompanying text reports for each combination of columns assigned *x* and *y* modeling roles.

A mosaic chart has side-by-side divided bars for each level of its *x* variable. The bars are divided into segments proportional to each discrete level (value) of the *y* variable. The mosaic chart in Figure 6.4 shows the relationship of marital status to the manufacturing country.

The width of each bar is proportional to the sample size. When the lines dividing the bars align horizontally, the response proportions are the same. When the lines are far apart, the response rates of the samples might be statistically different.

Figure 6.4 Mosaic Plot Axes

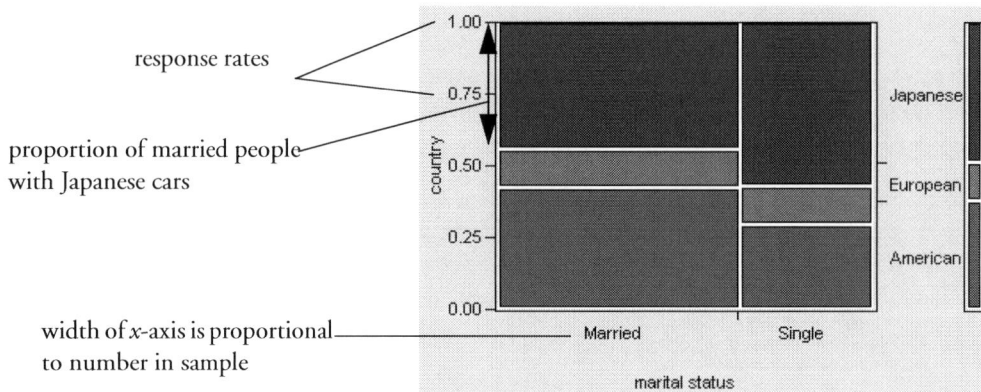

⊖ Scroll to see each *x* variable as it relates to manufacturing country.

Sex and **country** do not appear to have any relationship at all. The proportion of automobiles from the three manufacturing countries is about the same for each sex.

The country by age group mosaic plot shows that the proportion of American car owners 30 years or over is only slightly greater than the proportion of American car owners under age 30.

The most significant relationship is seen between **marital status** and **country**. The mosaic plot, shown previously in Figure 6.4, and its supporting **Tests** table (Figure 6.5), suggest that married people are more likely than single people to own American cars.

The Likelihood Ratio and Pearson Chi-squared tests evaluate the relationship between an automobile's country of manufacture and the marital status of owner. If no relationship exists between **country** and **marital status**, a smaller Chi-squared value than the one computed in this survey would occur only seven times in 100 similar surveys.

Figure 6.5 Table of Statistical Tests for country By marital status

Tests			
N	**DF**	**-LogLike**	**RSquare (U)**
303	2	2.5700361	0.0086

Test	**ChiSquare**	**Prob>ChiSq**
Likelihood Ratio	5.140	0.0765
Pearson	5.081	0.0788

These statistical results reveal that American auto manufacturers might want to direct advertising plans toward married couples.

🖑 Scroll the report to see the relationship between size of car and each *x* variable (sex, marital status, and age).

The three mosaic plots indicate no relationship between car size and gender, marital status, or age group. This is seen numerically by looking at the Contingency Tables and the Tests tables beneath each of the mosaic plots. (See Figure 6.6.)

Note that by default, **Col%** and **Row%** also appear in the Contingency Tables. Right-click (hold the CONTROL key and click on the Macintosh) the table to access the Columns menu to turn columns on and off.

The Chi-squared values support the hypothesis that the purchase of large, medium, and small cars is not significantly different across the sex, marital status, and age group factor levels. The Chi-squared probabilities range from 0.06 to 0.30, so you should expect smaller Chi-squared values to occur six to 30 times in 100 similar surveys.

It probably makes no difference what size cars appear in advertisements.

Figure 6.6 Tables for Relationships with Size of Car

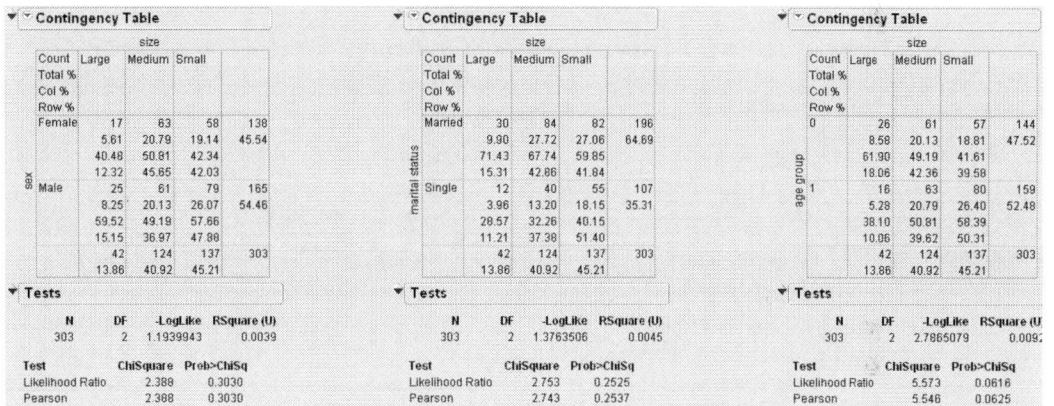

Contingency Table

Count / Total % / Col % / Row %		Large	Medium	Small	
sex	Female	17	63	58	138
		5.61	20.79	19.14	45.54
		40.48	50.81	42.34	
		12.32	45.65	42.03	
	Male	25	61	79	165
		8.25	20.13	26.07	54.46
		59.52	49.19	57.66	
		15.15	36.97	47.98	
		42	124	137	303
		13.86	40.92	45.21	

Tests

N	DF	-LogLike	RSquare (U)
303	2	1.1939943	0.0039

Test	ChiSquare	Prob>ChiSq
Likelihood Ratio	2.389	0.3030
Pearson	2.388	0.3030

Contingency Table

Count / Total % / Col % / Row %		Large	Medium	Small	
marital status	Married	30	84	82	196
		9.90	27.72	27.06	64.69
		71.43	67.74	59.85	
		15.31	42.86	41.84	
	Single	12	40	55	107
		3.96	13.20	18.15	35.31
		28.57	32.26	40.15	
		11.21	37.38	51.40	
		42	124	137	303
		13.86	40.92	45.21	

Tests

N	DF	-LogLike	RSquare (U)
303	2	1.3763506	0.0045

Test	ChiSquare	Prob>ChiSq
Likelihood Ratio	2.753	0.2525
Pearson	2.743	0.2537

Contingency Table

Count / Total % / Col % / Row %		Large	Medium	Small	
age group	0	26	61	57	144
		8.58	20.13	18.81	47.52
		61.90	49.19	41.61	
		18.06	42.36	39.58	
	1	16	63	80	159
		5.28	20.79	26.40	52.48
		38.10	50.81	58.39	
		10.06	39.62	50.31	
		42	124	137	303
		13.86	40.92	45.21	

Tests

N	DF	-LogLike	RSquare (U
303	2	2.7865079	0.009?

Test	ChiSquare	Prob>ChiSq
Likelihood Ratio	5.573	0.0616
Pearson	5.546	0.0625

The market survey categorizes cars based on both size and type, where a car's type is work, sporty, or family.

🖑 Scroll to see the plots that show the relationship between type of car and the three *x* variables.

The mosaic plots in Figure 6.7 show that the type of car varies for levels of marital status and age group. As perhaps expected, many of the cars owned by married people are family automobiles, while the largest proportion of cars owned by single people are sporty cars.

Figure 6.7 Reports for Type of Car and Marital Status and Age Group

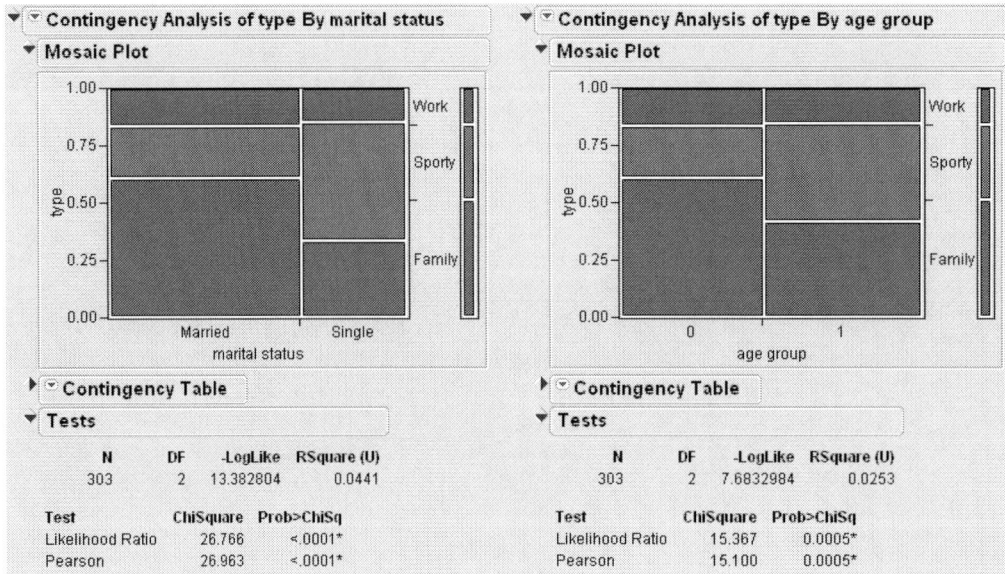

So, American automobile manufacturers might choose to focus advertisements toward married couples buying family-type automobiles.

It follows logically that a relationship between age group and type of car also exists because older people are more likely to be married. The graph to the right in Figure 6.7 shows graphically that the proportion of people over 30 years old who own family cars is much greater than those under 30. The small Chi-squared values support the significant difference in proportions. The Chi-squared values of 0.0005 mean that proportions as varied as these are expected to occur only five times in 1,000 similar surveys.

Chapter Summary

This chapter looked at relationships between categorical variables obtained from a survey. The survey recorded age, sex, marital status, and information about the type of automobile owned by a random sample of people in the same geographical area. The auto information included manufacturing country, size, and type of car. Car types were classified as **work**, **sporty**, and **family**. The question "Is the size of car, type of car, or manufacturing country related to the age, gender, or marital status of the owner?" was investigated.

The **Fit Y by X** command produced nine mosaic charts with supporting statistical summaries that show:

- No relationship between either sex or age and manufacturing country.
- A significant relationship between marital status and manufacturing country with married people more likely to own American cars than single people.

- No relationship between sex, age, or marital status and size of car.

- No relationship between sex and type of car.

- Significant relationships between marital status and type of car. As might be expected, married people over 30 years old were more likely to own family type cars than younger, single people.

The chapter "Contingency Tables Analysis" in the *JMP Statistics and Graphics Guide* discusses analyzing categorical data in more detail.

For more information about using the formula editor, see the chapter "Using the Formula Editor" in the *JMP User Guide*.

Chapter **7**

Regression and Curve Fitting
Visualizing Relationships

This lesson demonstrates the interactive regression capabilities of JMP.

The data is from Eppright *et al* (1972) as reported in Eubank (1988, p. 272). The study subjects are young males. The variables in the data table are age (in months) and the ratio of weight to height. A third variable classifies the subjects into two groups based on age. The goal is to describe and model the growth pattern of subjects for the age range given in the data table.

Objectives

- Use the **Fit Y by X** command to fit least squares lines to continuous data.
- Fit polynomial curves and cubic splines to the data set and explore their goodness of fit.
- Journal and save analysis results.
- Use the **Group By** command to fit different lines to certain groups of data.

Contents

Look Before You Leap

The first step is to become familiar with the data. Begin by reviewing the data to determine the best way to proceed with the regression.

Open a JMP File

🖱 When you installed JMP, a folder named Sample Data was also installed. In that folder is a file named Growth.jmp. Open Growth.jmp.

A partial listing of the Growth.jmp data table is shown here.

There are 2 columns and 72 rows. The ratio column contains the average weight-to-height ratio for each age group in the study. The age groups range from 0.5 to 71.5 months.

The modeling type for each column is shown to the left of the variable name in the columns panel. Both columns have a continuous modeling type (🔺), as needed for a regression analysis in JMP.

	ratio	age
1	0.46	0.5
2	0.47	1.5
3	0.56	2.5
4	0.61	3.5
5	0.61	4.5
6	0.67	5.5
7	0.68	6.5
8	0.78	7.5
9	0.69	8.5
10	0.74	9.5
11	0.77	10.5
12	0.78	11.5
13	0.75	12.5
14	0.80	13.5

Growth — Notes Eubanks (1988) Splin — Bivariate — Columns (2/0): ratio, age — Rows: All rows 72, Selected 0, Excluded 0, Hidden 0, Labelled 0

The purpose of the analysis is to determine whether the ratio values are related to (or dependent on) the age values.

Select an Analysis

To fit regression curves:

🖱 Select **Analyze > Fit Y by X**.

The **Fit Y by X** analysis does four types of analyses, depending on the modeling type of the variable:

- Regression analysis when both x and y have continuous values, as in this example.
- Categorical analysis when both x and y have nominal or ordinal values.
- Analysis of variance when x is nominal and y has continuous values.
- Logistic regression when x is continuous and y has nominal or ordinal values.

Choose Variable Roles

The **Fit Y by X** command first displays the Fit Y by X window. *Y* identifies a response or dependent variable and *x* identifies a classification or independent variable. To choose variable roles:

🖱 Highlight ratio and click **Y, Response**.

🖱 Highlight age and click **X, Factor**, as shown in Figure 7.1.

🖱 Click **OK**.

Figure 7.1 The Fit Y By X Window

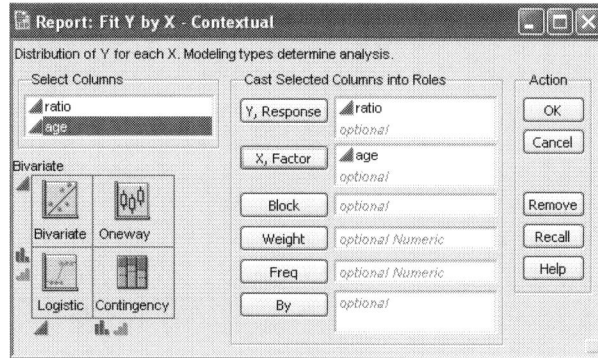

Now investigate if the ratio of weight to height is a function of age.

Fitting Models to Continuous Data

The scatterplot shown in Figure 7.2 is the result of the Fit Y by X analysis. It is easy to see that the growth pattern is not random. A straight line regression is a good baseline fit to compare with other regression curves.

Figure 7.2 Scatterplot of ratio by age

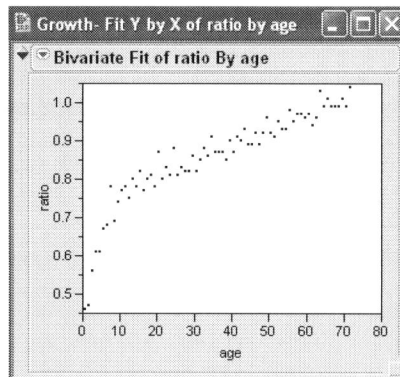

When clicked, the red triangle icon on the scatterplot title bar reveals a variety of fitting commands and additional display options. Options include **Show Points**, fitting commands, and other features. The **Show Points** command alternately hides or displays the points in the plot. Fitting options can be as simple as fitting a straight line or involved as drawing density ellipses. Fitting options can be used repeatedly to overlay different fits on the same scatterplot.

Begin with a simple line and try different techniques after inspecting the initial straight line regression fit.

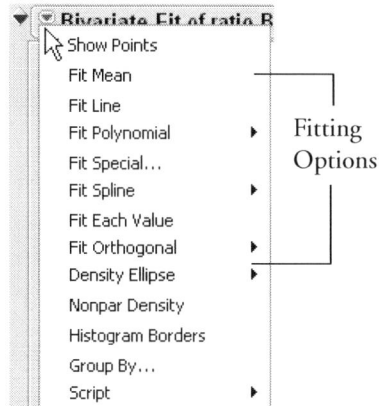

Fitting the Mean

🖑 Click the red triangle icon and select **Fit Mean**.

This is the baseline fit that hypothesizes that there is no relationship between x and y. All other fits compare to this fit. Since the Fit Mean table is closed by default:

🖑 Click the disclosure button by Fit Mean (◆ ◆ on Windows and Linux and ▶ ▼ on the Macintosh).

This displays a report that shows:

- The sample mean (arithmetic average) of the response variable.

- The standard deviation of the response variable.

- The standard error of the response mean.

- The error sum of squares for the simple mean model.

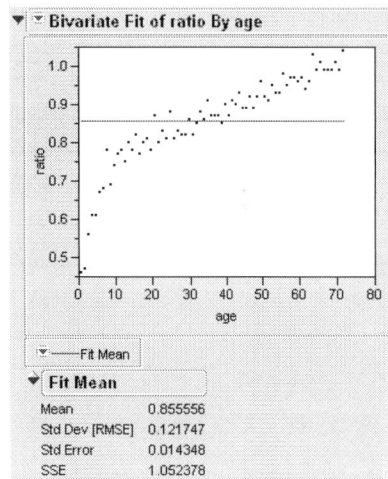

Fitting a Line

To fit a simple regression line through the data points:

🖑 Click the red triangle icon in the title bar and select **Fit Line**.

The regression line minimizes the sum of squared distances from each point to the line of fit. Because of this property, it is sometimes referred to as the *line of best fit*.

Red triangle icons

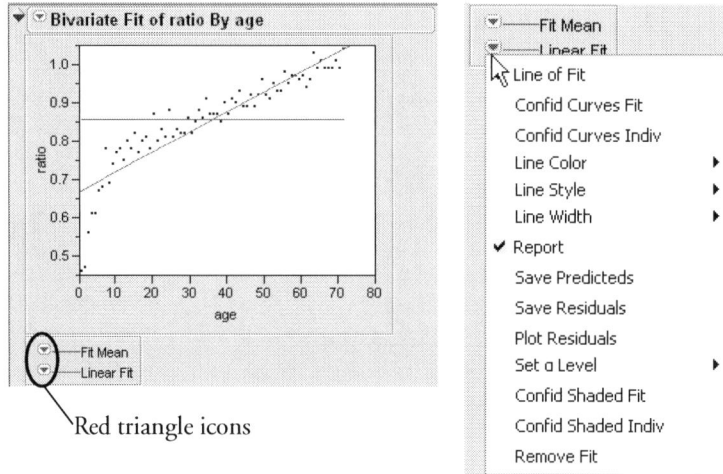

Each time a fit is selected from the red triangle icon, the regression equation and another red triangle icon for that fit show beneath the scatterplot, as shown here.

Click the red triangle icon to reveal commands that show confidence curves and give the ability to save predicted and residual values as new data table columns. The **Save Predicteds** command saves the prediction equation for the fit with the new column of predicted values. The fit can be removed from the scatterplot at any time with the **Remove Fit** command.

Understanding the Summary of Fit Table

Clicking the red triangle icon in the title bar and selecting **Fit Line** produces a Summary of Fit table, which summarizes the linear fit.

- **Rsquare (R^2)** quantifies the proportion of total variation in the growth ratios accounted for by fitting the regression line.
- **Rsquare Adj** adjusts R^2 to make it more comparable over models with different numbers of parameters.
- **Root Mean Square Error (RMSE)** is a measure of the variation in the ratio values that is attributable to different people rather than to different ages.
- **Mean of Response** is the arithmetic mean (average) of the ratio values.
- **Observations** is the total number of nonmissing values.

Note: The first line of the report is the regression equation, which is editable.

Understanding the Analysis of Variance Table

In addition to producing a Summary of Fit table, clicking the red triangle icon and selecting **Fit Line** produces an Analysis of Variance table.

▼ **Analysis of Variance**

Source	DF	Sum of Squares	Mean Square	F Ratio
Model	1	0.8656172	0.865617	324.4433
Error	70	0.1867605	0.002668	Prob > F
C. Total	71	1.0523778		<.0001*

The elements of the table give an indication of how well the straight line fits the data points:

- **Source**   identifies the sources of variation in the growth ratio values (Model, Error, and C. Total).
- **DF**   records the associated degrees of freedom for each source of variation.
- **Sum of Squares**   (SS for short) quantifies the variation associated with each variation source. The C. Total SS is the corrected total SS computed from all the ratio values. It divides (partitions) into the SS for Model and SS for Error. The Model SS is the amount of the total variation in the ratio scores explained by fitting a straight line to the data. The Error SS is the remaining or unexplained variation.
- **Mean Square**   lists the Sum of Squares divided by its associated degrees of freedom (DF) for Model and Error.
- **F Ratio**   is the regression (Model) mean square divided by the Error mean square.
- **Prob > F**   is the probability of a greater *F*-value occurring if the ratio values differed only because of different subjects rather than because the subjects are different ages.

In this example, the significance of the *F*-value is 0.0001, which strongly indicates that the linear fit to the weight/height growth pattern is significantly better than the horizontal line that fits the sample mean to the data.

Understanding the Parameter Estimates Table

In addition to producing a Line of Fit table and an Analysis of Variance table, clicking the red triangle icon and selecting **Fit Line** produces a Parameter Estimates table.

▼ **Parameter Estimates**

Term	Estimate	Std Error	t Ratio	Prob>|t|
Intercept	0.6656231	0.012176	54.67	<.0001*
age	0.0052759	0.000293	18.01	<.0001*

- **Term**   lists the parameter terms in the regression model.
- **Estimate**   lists estimates of the coefficients in the regression line equation.
- **Std Error**   lists estimates of the standard error of the parameters.
- **t Ratio**   is the parameter estimate divided by its standard error.
- **Prob > |t|**   is the probability of a greater absolute *t-value* occurring by chance alone if the parameter has no effect in the model.

The significant *F*-ratio in the Analysis of Variance table tells the student that the regression line fits significantly better than the horizontal line at the mean (the simple mean model). However, while the regression line looks like a good fit for age groups above seven months, it does not describe the data well for ages younger than seven months.

Excluding Points

Because the low-age points are the trouble spots for the linear fit, remove them from the analysis and try fitting the model to the remaining values.

To highlight these outliers and exclude them from the analysis:

✏ Select the lasso tool from **Tools** menu or toolbar.

✏ Drag the lasso around the points to be excluded.

✏ Select **Rows > Exclude/Unexclude** to exclude the selected points.

✏ Right-click the selected points, select **Row Markers**, and select **X** to assign the **X** marker to the excluded points.

✏ Click the red triangle icon in the title bar and choose the **Fit Line** command again to see the results of excluding the low-age points.

The scatterplot shown here has both regression lines. The low-age points still show on the plot but are not included in the second regression line's computation.

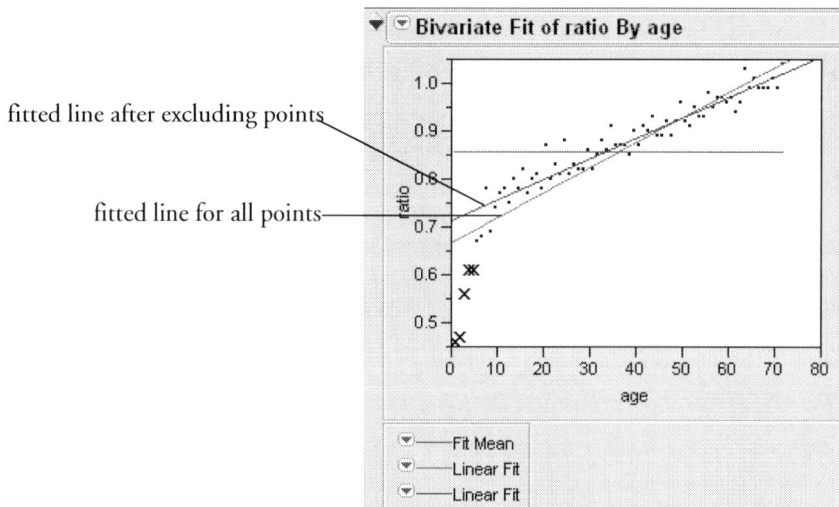

Journaling JMP Results

After completing this part of the exploratory regression analysis:

✏ Choose **Edit > Journal**.

The first time the **Journal** command is selected during a JMP session, a journal window opens and is filled with the graphs and tables from the report window.

The open journal file contains all reports from the active report window. Plots can be resized, opened, or closed, as can outlines. This allows for printing of certain parts of the report.

⟐ Choose **Save As** from the **File** menu to save the journal.

The window similar to the one shown here prompts for a filename, and appends .jrn to the filename to identify the file type.

Leaving a journal file open causes each subsequent use of the **Journal** command to append results in the active window at the end of the journal contents.

⟐ Name the journal Regression Results.

⟐ On Windows, change the **Save as type** to **RTF Files (*.RTF)** and click **Save**.

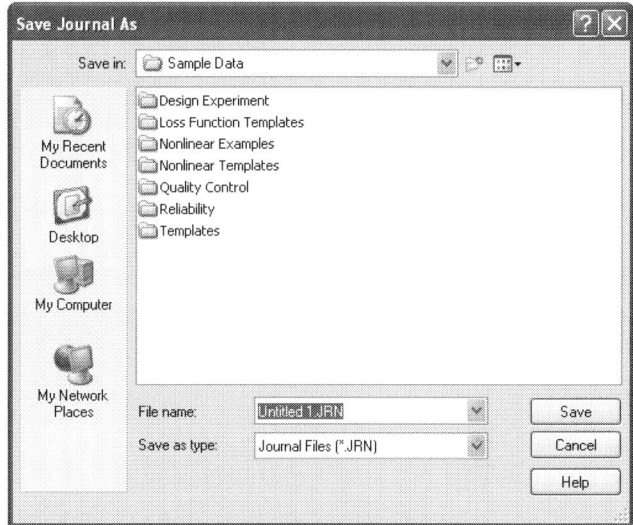

On Linux, change the **Save as type** option to ***.rtf (Rich Text File)** and click **Finished**.

On the Macintosh, select **Export** from the **File** menu instead of **Save As**. Select **RTF** and click **Next**. Click **Export**.

⟐ Navigate to the file's directory on your system and open the file.

The file should open in your default word processor as shown here. Note that the graphics are saved as graphics, and the reports are saved as text tables.

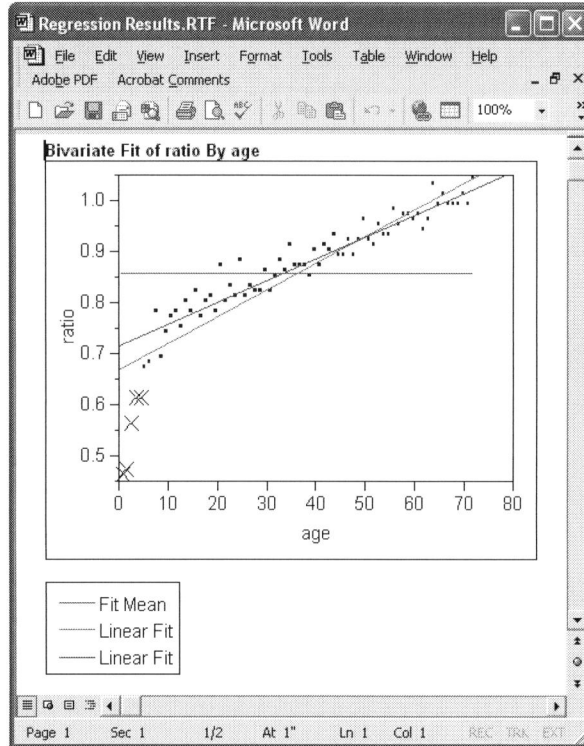

Examining a Polynomial Fit (Linear Regression)

Now, examine a polynomial fit for comparison to the linear fit. A linear regression is simply a polynomial of degree 1.

🖰 Click the Bivariate Fit of ratio By age report window to assure it is active.

🖰 To remove the row exclusions and markers, select **Rows > Clear Row States**.

🖰 Click the red triangle icon in the title bar, select **Fit Polynomial > 2,quadratic**, which allows the fit to have curvature.

🖰 Remove the line of fit that excluded the lower age groups by clicking the *second* Linear Fit (modified) regression line's red triangle icon and selecting **Remove Fit**.

🖰 Remove the Fit Mean results so that only the polynomial fit and the line fit to all the data points remain. Select **Remove Fit** from the Fit Mean red triangle icon.

🖰 Click the red triangle icon in the title bar and select **Fit Polynomial > 3,cubic** to overlay a polynomial curve of degree 3 on the scatterplot.

🖰 Again select the **Edit > Journal** command to append these results to the existing journal.

Figure 7.3 Comparison of Linear Fit and Polynomial Fits of Degree 2 and 3

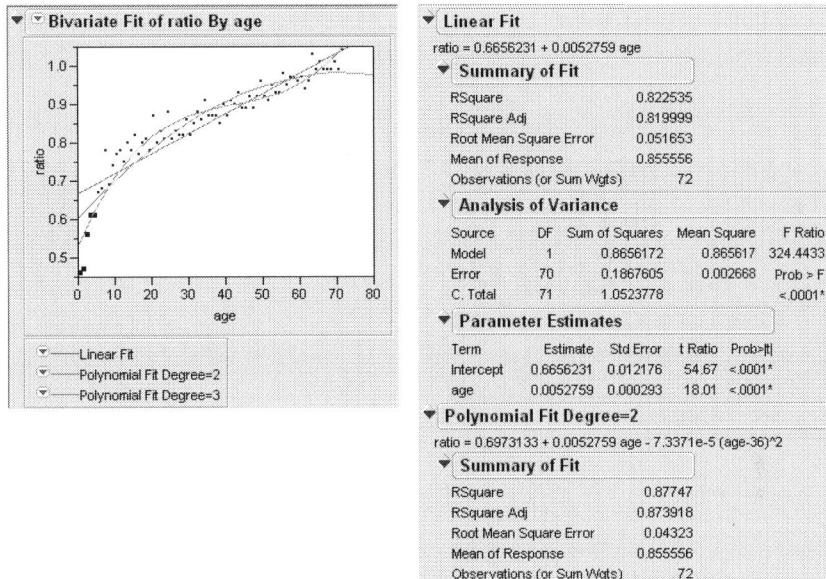

The tables show the R^2 value from the Summary of Fit tables for the linear fit, the second degree polynomial fit, and the third degree fit. As polynomial terms are added to the model, the regression curve appears to fit the data better. (See Figure 7.3 for the graph and some of the tables.)

Fitting a Spline

Even the polynomial fit of degree 3 does not quite reach the outlying points of the very young subjects. A free-form function that acts as if it smooths the data, such as a smoothing *spline*, might be better.

🖑 Use the **Remove Fit** command on both polynomial fits, so that only the first linear regression line shows on the scatterplot.

🖑 Click the red triangle icon on the title bar and select **Fit Spline** three times, with lambda values of 10, 1,000, and 100,000.

Lambda is a tuning factor that determines the flexibility of the spline. The **Fit Spline** command submenu (shown to the left in Figure 7.4) lists lambda values. The three new fits are overlaid on the scatterplot.

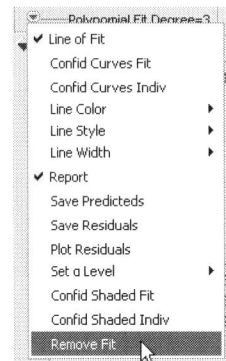

Figure 7.4 Comparison of Spline Fits

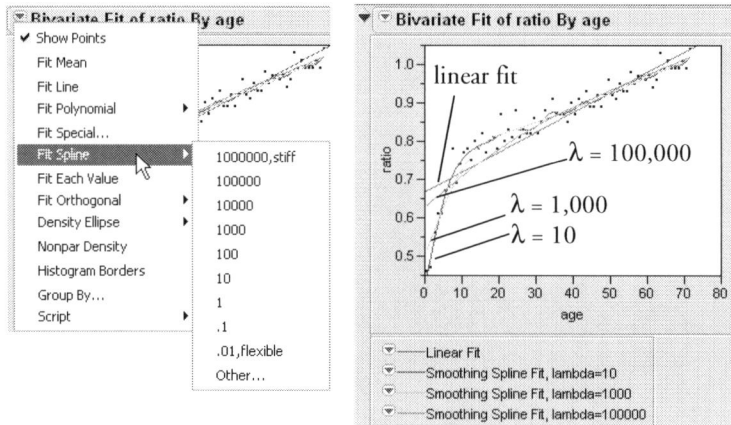

By inspecting the plot, see that the lambda = 10 curve is too flexible and therefore local error has too great an effect on it. The lambda = 100,000 curve is too stiff. It is so straight that it does not reach down to model the lower ages closely. However, the lambda of 1,000 curve fits well. Its shape is not influenced by local errors, and it appears to fit the data smoothly.

If a report of these results is needed, journal these results.

🖑 Select **Edit > Journal**.

The **Journal** command appends the scatterplot with spline fits and text reports to the open journal file. After journaling the final analyses, the following draft notes about the spline-fitting technique can be added at the bottom of the journal window.

🖑 Select the annotate tool from the toolbar.

🖑 Click and drag a large box at the bottom of the report.

🖑 Add the following text to the box.

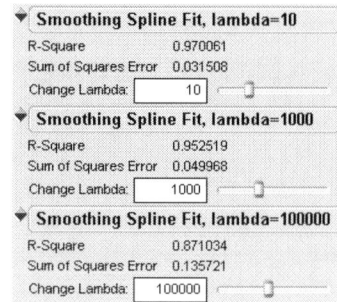

> ```
> "This fitting technique applies a cubic polynomial to the interval
> between points; the polynomial is joined such that the curve meets at the
> same point with the same slope to form a continuous and smooth curve. A
> small enough lambda could make such a curve go through every point, which
> would model the error, not the mean. A moderate lambda value forces the
> curve to be smoother, or in other words, less curved. This is
> accomplished by adding a curvature penalty to the optimization that
> minimizes the sum of squares error."
> ```

By comparing various regression fits, notice that both the polynomial fits and the spline fit with moderate flexibility best describe the data. These models show that infants grow most rapidly during the first months of life and that growth rate decreases significantly at approximately 12 months.

Fitting By Groups

"Excluding Points," p. 88 in this chapter, shows how to overlay a linear fit for the whole sample with a linear fit for children over the age of seven months. Carry this idea one step further with overlay fits to compare children under the age of one year with children over one year.

🚪 In the Growth.jmp data table, create a new column called group to act as a grouping variable. Right-click in the new column area of the data table and select **New Column** from the resulting menu. Write the column name and click **OK**.

🚪 Right-click in the Group column (hold the CONTROL key and click on the Macintosh) and select **Formula**.

Now enter the formula shown in Figure 7.5:

🚪 Click **Conditional** in the function selector list and select the **If** function.

🚪 The expression term, denoted expr, is highlighted.

🚪 Choose **a < b** from the Comparison functions.

🚪 The left side of the comparison clause is highlighted. Click age in the column selection list.

🚪 Enter **12** for the numeric comparison.

🚪 Double-click the term denoted then clause.

🚪 Enter **"Babies"** (in double quotes because this column is a character variable).

🚪 Double-click the term denoted else clause.

🚪 Enter **"Toddlers"** (with double quotes).

🚪 Click **Apply** and then **OK**.

This assigns the value Babies to each child less than 12 months old, and Toddlers to children who are 12 months or older.

Figure 7.5 Computed Age Grouping Variable

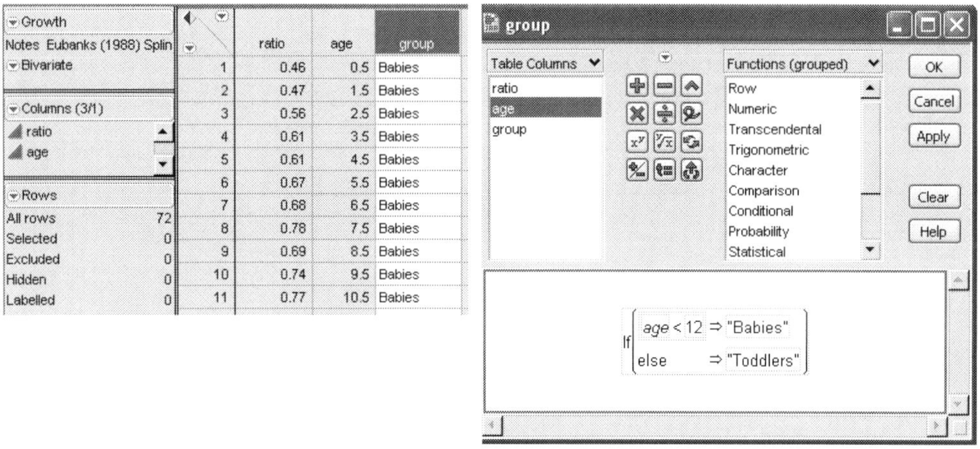

🚪 Click the Bivariate report to make it the active window.

🖑 Clear the Smoothing Spline fits still showing, such as those seen in Figure 7.4, using each fit's **Remove Fit** command. Click the red triangle icon for all three smoothing spline fits and select **Remove Fit** for each one.

🖑 Click the red triangle icon and select **Group By** to display the window shown here.

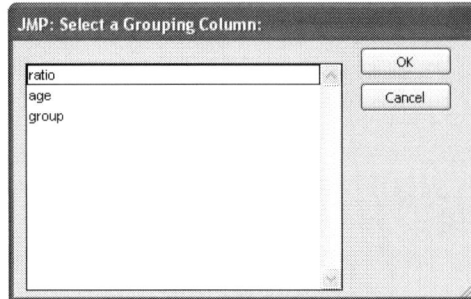

🖑 Select group, the newly created grouping variable, and click **OK**.

🖑 Choose the **Fit Line** command.

With a grouping variable (group) in effect, the overlaid regression lines shown in Figure 7.6 appear automatically. The points that correspond to each regression give a dramatic visualization of the steep growth rate for babies during the first year of life compared to the more moderate growth rate of toddlers and small children age one to five years.

Figure 7.6 Regression Lines for Levels of a Grouping Variable

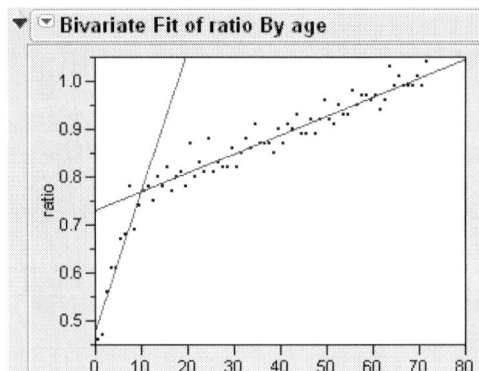

Chapter Summary

To analyze some bivariate data, the **Fit Y by X** command was used to examine a variety of regression model fits. The task was to model and describe the growth pattern of subjects over a range of ages. You measured growth using the ratio of weight to height and accomplished this task by:

- Fitting mean to use as a baseline comparison to other regression models and evaluate the fit using statistical text reports.

- Fitting a straight line as a first guess for a model.
- Excluding outliers and again fitting a straight line to compare the R^2 values given by the Summary of Fit tables for both lines.
- Fitting second and third degree polynomials to see whether they model the growth pattern more realistically.
- Fitting smoothing splines with lambda values of 10, 1,000, and 100,000 and comparing them with each other and with the linear fit.
- Clicking the red triangle icon and selecting the grouping facility (**Group By**) to compare growth rates of babies under the age of one year with toddlers from age one to five years.
- Using the **Journal** command to append each of these regression reports and graphs to a journal file.

7 Regression

Chapter 8

A Factorial Analysis
Designed Modeling

This lesson examines two treatments of popcorn. The plain, everyday type has been around for years, but researchers claim to have discovered a special treatment of corn kernels. This new process supposedly increases the popcorn yield as measured by popcorn volume from a given measure of kernels.

Is this true? If so, how much is the increase? Are these increases the same for all groups of conditions? The special treatment raises the cost of the popcorn, so the increase in yield must be significant enough to warrant the higher costs.

The popcorn data used in this chapter and for examples in the *JMP User Guide* and the *JMP Statistics and Graphics Guide* are artificial, but the experiment was inspired by experimental data reported in Box, Hunter, and Hunter (1978).

Objectives

- Learn techniques to analyze a designed factorial experiment using the **Fit Model** command.
- Evaluate and interpret effects using interactive graphic tools.
- Examine supporting text reports.
- Evaluate the significance of interaction effects using interaction plots.
- Save a model's predicted values for each observation.

Contents

Look Before You Leap

The popcorn yield data are the result of a designed experiment. The same amounts of different types of corn were methodically popped under different conditions. First, look at the data to review the results of the popcorn experiment.

Open a Data Table

🖑 When you installed JMP, a folder named Sample Data was also installed. In that folder is a file named Popcorn.jmp. Open Popcorn.jmp.

The Popcorn data table displays in spreadsheet form as shown here.

For the experiment, the corn was popped under controlled conditions. Plain popcorn and specially treated gourmet popcorn were each popped in large or small amounts of oil and in large or small batches. Two trials were done for both types of corn under all popping conditions.

	popcorn	oil amt	batch	yield	trial
1	plain	little	large	8.2	1
2	gourmet	little	large	8.6	1
3	plain	lots	large	10.4	1
4	gourmet	lots	large	9.2	1
5	plain	little	small	9.9	1
6	gourmet	little	small	12.1	1
7	plain	lots	small	10.6	1
8	gourmet	lots	small	18.0	1
9	plain	little	large	8.8	2
10	gourmet	little	large	8.2	2

This experimental design is called a *factorial design*. The experiment has three factors, usually called *main effects*, which are:

- Type of popcorn (plain or gourmet)
- Amount of cooking oil (little or lots)
- Cooking batch size (large or small)

What Questions Can Be Answered?

The appropriate statistical analysis for a factorial design addresses the following questions about the main effects:

- Is there an overall difference in yield between plain and gourmet popcorn?
- Is there an overall difference in yield between cooking in lots of oil instead of a small amount of oil?
- What is the difference in yield between cooking several small batches instead of one large batch?

Analysis of a factorial experiment also provides information about the *interaction* between the main effects as addressed by the following questions:

- Does the amount of cooking oil have the same effect on both types of popcorn? In other words, is there an interaction effect between popcorn type and amount of cooking oil used?
- Is there an interaction effect between batch size and type of popcorn?
- Is there an interaction effect between batch size and amount of oil used?
- Are there interaction effects among the three main effects?

The Fit Model Window

🖰 Select **Analyze > Fit Model**.

The **Fit Model** command lets you specify and analyze complex models like the factorial design in this experiment.

The **Fit Model** command displays the Fit Model window shown in Figure 8.1. This window is used to define the type of model, the model response variable, and model effects.

To specify the factorial model:

🖰 Select yield from the **Select Columns** list.

🖰 Click the **Y** button.

🖰 Select popcorn, oil amt, and batch from the **Select Columns** list.

🖰 Click the **Macros** button and select **Full Factorial**. This adds all main effects and interactions (*crossed effects*) to the **Construct Model Effects** list (Figure 8.1).

🖰 Further tailor the model by adding effects or removing unwanted effects with the **Add** and **Remove** buttons. In this case, remove the three-way interaction term by highlighting popcorn*oil amt*batch and clicking **Remove**.

Figure 8.1 Fit Model Window

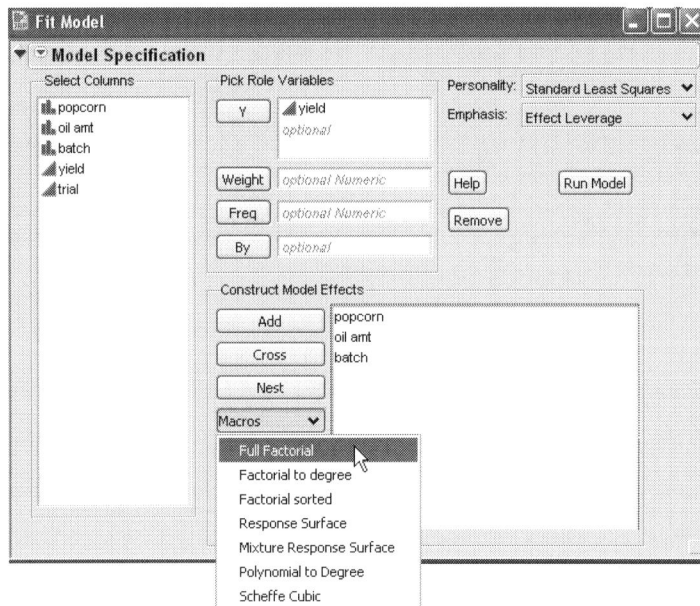

🖰 Select **Effect Leverage** from the box beside **Emphasis** at the top right of the Fit Model window.

🖰 Click **Run Model** to estimate the model parameters and view the results.

Graphical Display: Leverage Plots

The **Fit Model** command graphically displays the whole model and each model effect as the *leverage plots* shown in Figure 8.2 through Figure 8.5. It is possible to tell at a glance whether the factorial model explains the popcorn data and which factors are most influential.

The Whole Model plot to the left in Figure 8.2 shows actual yield by predicted yield values with a regression line and 95% confidence curves. The regression line and the 95% confidence curves cross the sample mean (the horizontal line), which show that the whole factorial model (all effects together) explains a significant proportion of the variation in popcorn yield.

There is also a significant difference in yield between the two types of popcorn, as shown in the right-hand leverage plot for the popcorn main effect. The small *p*-values beneath the plots quantify the significant model fit and popcorn effect.

Figure 8.2 Leverage Plots of Actual by Predicted and of **Popcorn** Effect

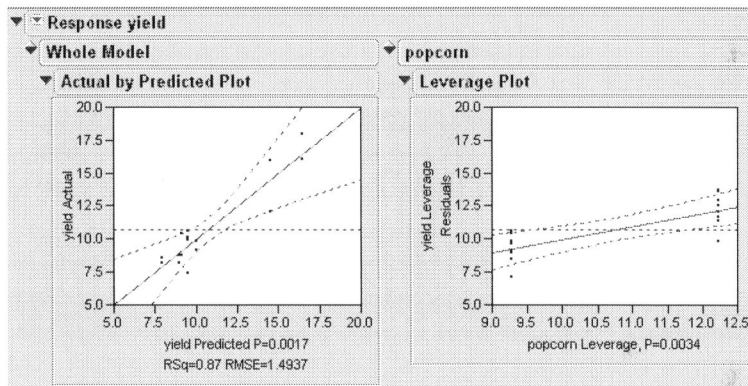

In Figure 8.3, the confidence curves for oil amt and the popcorn*oil amt interaction do not cross the horizontal mean line (rather, they encompass the mean line). This shows that neither of these factors significantly affected popcorn yield.

Figure 8.3 Leverage Plots for the Oil Amt and Its Interaction with Popcorn

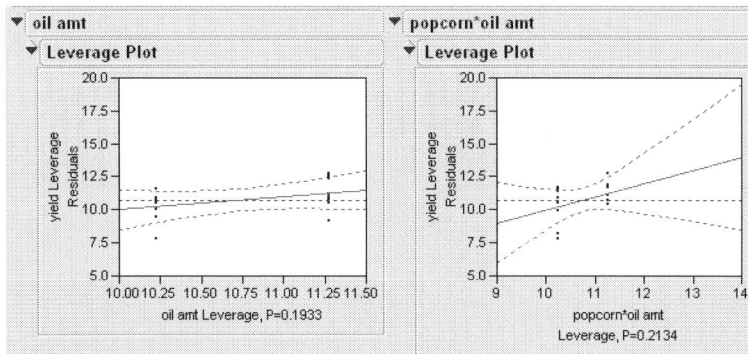

The leverage plots in Figure 8.4 show that the batch size effect (batch) and the interaction between popcorn type and batch size (popcorn*batch) are significant effects. This means that the size of the

batch makes a difference in the popcorn yield. Furthermore, the significant interaction means that batch size affects each type of popcorn differently.

Figure 8.4 Leverage Plots for Batch and Its Interaction with Popcorn

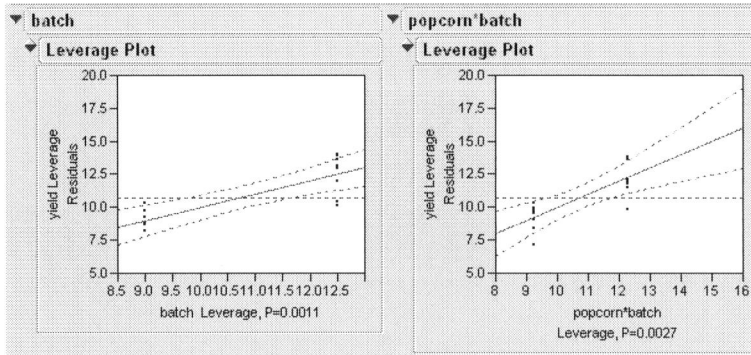

The two leverage plots shown in Figure 8.5 show that there is no significant interaction between amount of oil and batch size.

Figure 8.5 Leverage Plots for Other Interaction Effects

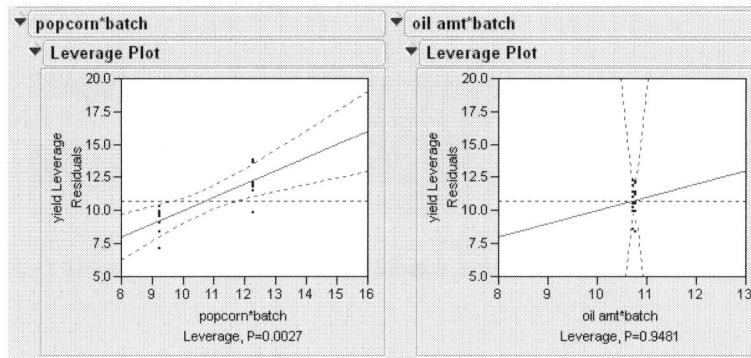

For more information about interpretation of leverage plots, see the chapters "Understanding JMP Analyses" and "Standard Least Squares: Introduction" and the appendix "Statistical Details" of the *JMP Statistics and Graphics Guide*.

Quantify Results: Statistical Reports

Because oil amt and its interactions with other effects are not significant, fit the popcorn data again without these effects. The new model should have the significant factors (type of popcorn, batch size), and their interaction term. This approach condenses the statistical reports that show estimates of yield under the different conditions of interest. Use the same **Fit Model** command as before.

🖱 Click the Fit Model window to make it the active window.

⏁ If the window is closed, click the red triangle icon on the report and select **Script > Redo Analysis** to open a new Fit Model window.

Use the following method to specify the two-factor model:

⏁ From the full factorial model, select unwanted effects listed in the **Construct Model Effects** box and click **Remove**.

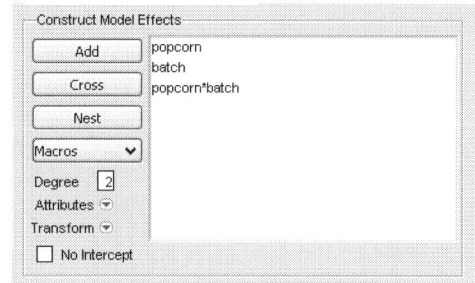

⏁ Click **Run Model**.

```
┌─ Construct Model Effects ──────────────────┐
│  ┌─────────────┐   popcorn                 │
│  │    Add      │   batch                    │
│  └─────────────┘   popcorn*batch            │
│  ┌─────────────┐                            │
│  │   Cross     │                            │
│  └─────────────┘                            │
│  ┌─────────────┐                            │
│  │   Nest      │                            │
│  └─────────────┘                            │
│  │Macros    ▼ │                             │
│                                             │
│  Degree   [2]                               │
│  Attributes ▼                               │
│  Transform ▼                                │
│  ☐ No Intercept                             │
└─────────────────────────────────────────────┘
```

Analysis of Variance

The whole model leverage plot in Figure 8.6 shows that the two-factor model describes the popcorn experiment well. Examine the tables that accompany the whole model leverage plot.

The Analysis of Variance table (in Figure 8.6) that accompanies the whole model leverage plot quantifies the analysis results. It lists the *partitioning* of the total variation of the sample into components. The ratio of the Mean Square components forms an F statistic that evaluates the effectiveness of the model fit. If the probability associated with the F-ratio is small, then the analysis of variance model fits better statistically than the simple model that contains only the overall response mean.

Figure 8.6 Analysis of Variance for the Two-Factor Whole Model

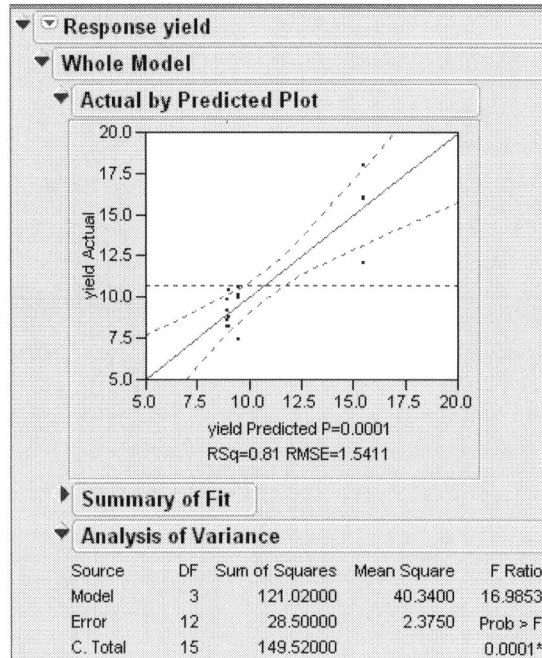

The Analysis of Variance table shows these quantities:

- **Source**   identifies the sources of variation in the popcorn yield values (Model, Error, and C. Total).
- **DF**   records the degrees of freedom for each source of variation.
- **Sum of Squares**   (SS for short) quantifies the variation in yield. C. Total is the corrected total SS. It is divided (partitioned) into the SS for Model and SS for Error. The SS for Model is the variation in the yield explained by the analysis of variance model, which hypothesizes that the model factors have a significant effect. The SS for Error is the remaining or unexplained variation.
- **Mean Square**   is a sum of squares divided by its associated degrees of freedom (DF).
- **F Ratio**   is the model mean square divided by the error mean square.
- **Prob > F**   is the probability of a greater F ratio occurring if the variation in popcorn yield resulted from chance alone rather than from the model effects.

In this example, the p-value (**Prob > F**) is 0.0001. JMP indicates a significant p-value by placing an asterisk beside it. The low value of this p-value implies that the difference found in the popcorn yield produced by this experiment is expected only 1 time in 10,000 similar trials if the model factors do not affect the popcorn yield.

Summary Reports For The Whole Model

Other tables in the Fit Model report provide statistical summaries. The Summary of Fit table shows the numeric summaries of the response for the factorial model:

- **Rsquare (R²)** of 0.809 tells the scientist that the two-factor model explains nearly 81% of the variation in the data.

- **Rsquare Adj** adjusts R^2 to make it more comparable over models with different numbers of parameters.

- **Root Mean Square Error** (sometimes called the RMSE) is a measure of the variation in the yield scores that can be attributed to random error rather than differences in the model's factors.

- **Mean of Response** is the mean (average) of the yield scores.

- **Observations** is the total number of recorded scores.

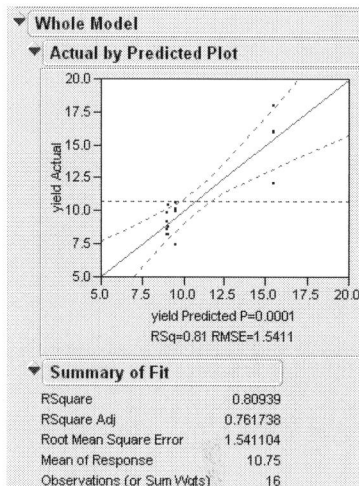

Whole Model

Actual by Predicted Plot

yield Predicted P=0.0001
RSq=0.81 RMSE=1.5411

Summary of Fit

RSquare	0.80939
RSquare Adj	0.761738
Root Mean Square Error	1.541104
Mean of Response	10.75
Observations (or Sum Wgts)	16

The *F* test probabilities in the Effect Test table tell the scientist that all model effects explain a significant proportion of the total variation. JMP indicates a significant *F* value by placing an asterisk beside it. There is also a table that gives the parameter estimates for the model.

Effect Tests

Source	Nparm	DF	Sum of Squares	F Ratio	Prob > F
popcorn	1	1	34.810000	14.6568	0.0024*
batch	1	1	49.000000	20.6316	0.0007*
popcorn*batch	1	1	37.210000	15.6674	0.0019*

Summary Reports for Effects

Now look at the summary tables for each effect in the model. The tables for the main effects are shown here. The Least Squares Means table lists the least squares means and standard errors for each level of the model factors, without considering the interaction between them. In this balanced example, the least squares means are simply the sample means of each factor level.

popcorn
 ▶ Leverage Plot
 Least Squares Means Table

Level	Least Sq Mean	Std Error	Mean
gourmet	12.225000	0.54486237	12.2250
plain	9.275000	0.54486237	9.2750

batch
 ▶ Leverage Plot
 Least Squares Means Table

Level	Least Sq Mean	Std Error	Mean
large	9.000000	0.54486237	9.0000
small	12.500000	0.54486237	12.5000

The nature of the interaction is important in the interpretation of the popcorn experiment. To examine the significant popcorn*batch interaction,

🖱 Click the red triangle icon from the Response yield title bar and select **Factor Profiling > Interaction Plots**.

This command plots the least squares means for each combination of effect levels, as shown in Figure 8.7.

Figure 8.7 Interaction Plots

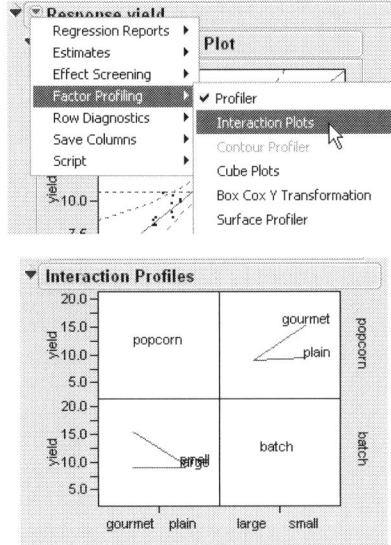

The Least Squares Means table for the **popcorn*batch** effect tells the whole story. Batch size makes no difference for the plain brand popcorn, but popping in small batches increases the yield in the new gourmet brand.

Because the factorial model with two-factors is a good prediction model, save the prediction formula.

🖱 Click the red triangle icon in the **Response yield** title bar and select **Save Columns > Prediction Formula**, as shown in Figure 8.8.

Figure 8.8 Prediction Formula for Popcorn Yield

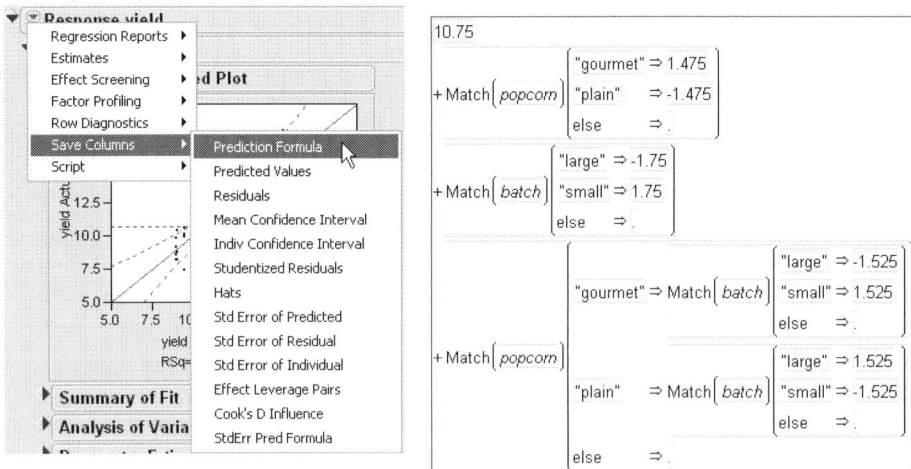

This command creates a new column in the Popcorn data table called Pred Formula yield that contains the predicted values for each experimental condition.

The prediction formula, shown at the bottom of Figure 8.8, becomes part of the column information. To see this formula:

🖑 Highlight the new column name (Pred Formula yield).

🖑 Select **Formula** from the **Cols** menu.

The prediction formula can be copied to the clipboard using standard cut and paste techniques.

Results show that popcorn should be packaged:

- in small packages so that the yield is good.

- in family size packages with smaller packets inside.

- in family size packages with popping instructions that clearly state the best batch size for good results.

Chapter Summary

In this chapter, a designed experiment evaluated the difference in yield between two types of popcorn. A three-factor factorial experimental design was the basis for popcorn popping trials. The results were analyzed by using the **Analyze > Fit Model** command. The following results were found:

- The leverage plots for the factorial analysis of three factors showed one main effect and its associated interactions to be insignificant.

- A more compact two-factor analysis with interaction adequately described the variation in yield for the popcorn trials.

- The interaction between the two main effects was significant. The Least Squares Means table for the interaction showed how the two types of popcorn behaved under different popping conditions.

The new, more expensive gourmet popcorn had better yield than the plain everyday type only if popped in small batches.

Chapter 9

Exploring Data
Finding Exceptions

Exploration is the search to find something new—the endeavor to make some discovery. For data analysis, exploratory study is often the most fruitful part of the analytical process because it is the most open to serendipity. Something noticed in a data set can be the seed of an important advance.

There are two important aspects of exploration:

- What is the pattern or shape of the data?
- Are there points unusually far away from the bulk of the data (outliers)?

When exploring data composed of many variables, the great challenge is dealing with this high dimensionality. There can be many variables that have interesting relationships, but it's hard to visualize the relationship of more than a few variables at a time.

Objectives

- Use graphical techniques to search for outliers in one, two, three, and higher dimensions.
- Perform a principal components analysis and examine it graphically.
- Examine outliers graphically using Mahalanobis distance.

Contents

Solubility Data

This lesson examines compounds for those with unusual solubility patterns in various solvents. When you installed JMP, a folder named Sample Data was also installed. In that folder is a file named Solubility.jmp. Data from an experiment by Koehler, Grigorus, and Dunn (1988) are in the Solubility.jmp file.

⟨ᵇ Open Solubility.jmp.

There are 72 compounds tested with six solvents, in columns called 1-Octanol, Ether, Chloroform, Benzene, Carbon Tetrachloride, and Hexane.

The Labels column in the table should serve as a label variable (Figure 9.1) so when you plot them, the compound names instead of row numbers identify points. Although this is already done for you in Solubility.jmp, you should know how to assign the label role to columns.

⟨ᵇ Select the columns.

⟨ᵇ Select **Cols > Label/Unlabel**.

or click the red triangle in the columns panel and ensure **Labels** is highlighted.

Figure 9.1 Solubility Data Table

		Labels	1-Octanol	Ether	Chloroform	Benzene	Carbon Tetrachloride	Hexane
	1	METHANOL	-0.770	-1.150	-1.260	-1.890	-2.100	-2.800
	2	ETHANOL	-0.310	-0.570	-0.850	-1.620	-1.400	-2.100
	3	PROPANOL	0.250	-0.020	-0.400	-0.700	-0.820	-1.520
	4	BUTANOL	0.880	0.890	0.450	-0.120	-0.400	-0.700
	5	PENTANOL	1.560	1.200	1.050	0.620	0.400	-0.400
	6	HEXANOL	2.030	1.800	1.690	1.300	0.990	0.460
	7	HEPTANOL	2.410	2.400	2.410	1.910	1.670	1.010
	8	ACETIC_ACID	-0.170	-0.340	-1.600	-2.260	-2.450	-3.060
	9	PROPIONICACID	0.330	0.270	-0.960	-1.350	-1.600	-2.140
	10	BUTYRICACID	0.790	0.610	-0.270	-0.960	-0.970	-1.760
	11	HEXANOICACID	1.920	1.950	1.150	0.300	0.570	-0.460
	12	PENTANOICACID	1.390	1.000	0.280	-0.100	-0.420	-1.000
	13	TRICHLOROACETICACI	1.330	1.210	-0.690	-1.300	-1.660	-2.630
	14	DICHLOROACETICACID	0.920	1.310	-0.890	-1.400	-2.310	-2.720

Columns panel (left side): Solubility, Notes Chemical compounds, Distribution, Spinning Plot, Multivariate, Columns (7/0), Labels, 1-Octanol, Ether, Chloroform, Benzene, Carbon Tetrachloride, Hexane, Rows, All rows 72, Selected 0

Use this menu to assign **Label** role to a column

There are six solvent variables, but there are no six-dimensional graphics. However, it is possible to look at six one-dimensional graphs, 15 two-dimensional graphs, and 20 three-dimensional plots. Using principal components, a representation of higher dimensions can be displayed.

One-Dimensional Views

The **Distribution** command helps you summarize data one column at a time. It does not show any relationships between variables, but the shape of the individual distributions helps identify the one-dimensional outliers.

To begin exploring the solubility data:

✏ Choose **Analyze > Distribution**.

✏ Select the six solubility columns and click the **Y, Columns** button.

✏ Click **OK**.

Their histograms, resized and trimmed of other output, are shown in Figure 9.2.

✏ Click any histogram bar.

That bar, and all other representations of that data, are highlighted in all related windows. To see how outlying values are distributed in the other histograms:

✏ Shift-click the outlying bars in each histogram.

This identifies the outlying rows in each single dimension.

✏ Use the **Rows > Markers** palette to assign the **X** marker to these selected rows.

The markers appear in the data table and in subsequent plots.

Figure 9.2 One-Dimensional Views

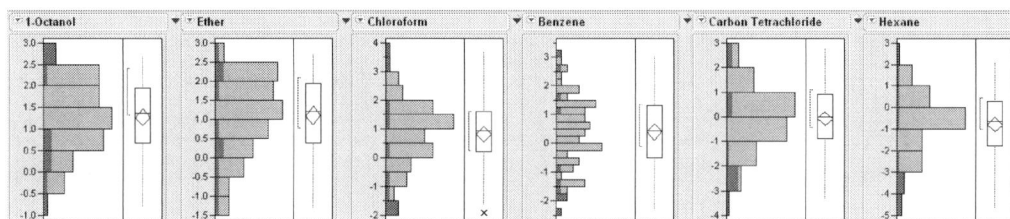

To create a new data table that contains only the outlying rows:

✏ Use the **Tables > Subset** command as shown here.

✏ Click **OK** to accept the default settings.

✏ Scroll through the new subset table to see the compound names of the one-dimensional outliers.

Two-Dimensional Views

✏ Return to Solubility.jmp.

✏ Select **Analyze > Multivariate Methods > Multivariate**.

✏ Highlight all the continuous columns in the table and click the **Y, Columns** button.

✏ Click **OK**.

This displays a correlation matrix and a scatterplot matrix of all 30 two-dimensional scatterplots (Figure 9.3).

The one-dimensional outliers appear as Xs in each scatterplot. Note in the scatterplot matrix that many of the variables appear to be correlated, as evidenced by the diagonal flattening of the normal bivariate density ellipses. There appear to be two groups of variables that correlate among themselves but are not very correlated with variables in the other group.

Figure 9.3 Two-Dimensional View

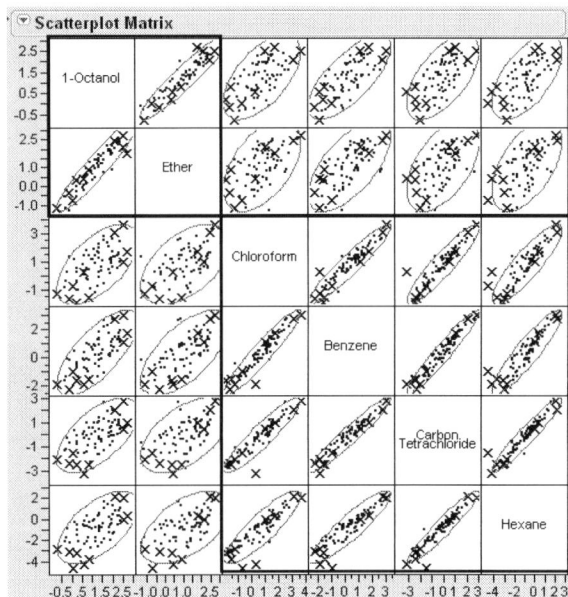

The variables Ether and 1-Octanol appear to make up one group, and the other group consists of the remaining four variables. These two groups are outlined on the scatterplot matrix shown in Figure 9.3.

Scan these plots looking for outliers (points that fall outside the bivariate ellipses) of a two-dimensional nature and identify them with square markers using the following steps.

- Double-click on **Selected** in the rows panel in the Solubility.jmp data table to clear your current selection.

- Shift-click each outlier.

- Select **Rows >Markers** and select the square marker from the palette.

Now, both one- and two-dimensional outliers are identified.

Three-Dimensional Views

To see points in three dimensions:

- Double-click on **Selected** in the rows panel in the Solubility.jmp data table to clear the row selection.

- Select **Graph > Scatterplot 3D**, which opens the Scatterplot 3D window.

- Add all six continuous variables to the **Y, Columns** list.

- Click **OK**.

- After the plot appears, change the drop-down lists below the plot to any combination of the three variables.

The goal is to look for points away from the point cloud for each combination of three variables. To aid in this search:

🖐 Rotate and examine each three-dimensional plot by dragging the plot with the mouse.

🖐 Hover over points to identify outliers.

Figure 9.4 shows two three-dimensional outlying points in the view of Ether by 1-Octanol by Benzene that hadn't been apparent before. To label them:

🖐 Shift-click these points.

🖐 Select **Rows > Label/Unlabel**.

Their labels, METHYLACETATE and ACETONE, appear on the plot.

Figure 9.4 Spotting Outliers in a Three-Dimensional View

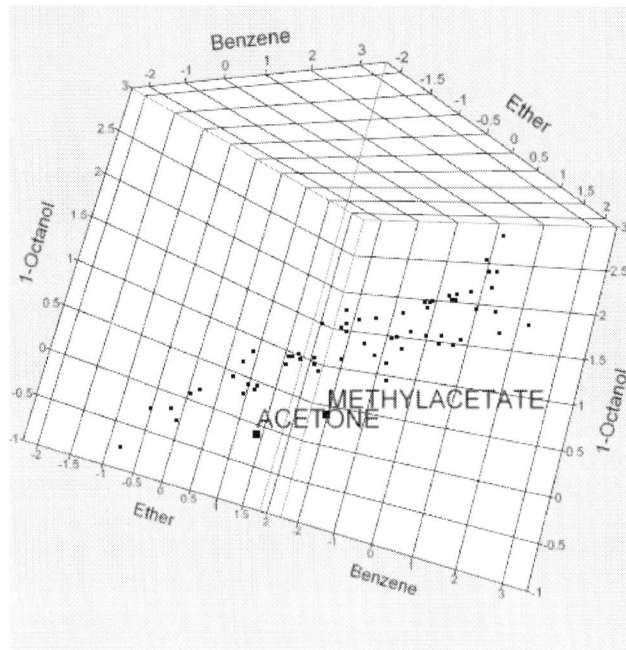

Principal Components and Biplots

Because many of the variables in the Solubility.jmp table are highly correlated, there is not a lot of scatter in six dimensions. The scatter is oriented in some directions but is flattened in other directions.

To illustrate this:

⍵ Remove the labels from METHYLACETATE and ACETONE by selecting **Rows > Label/ Unlabel**.

⍵ Select **Analyze > Multivariate Methods > Principal Components**.

Add only two highly correlated variables, in this case Ether and 1-Octanol.

⍵ Highlight Ether and 1-Octanol, select **Y, Columns** and click **OK**.

⍵ Click the red triangle icon in the Principal Components title bar and select **Spin Principal Components**.

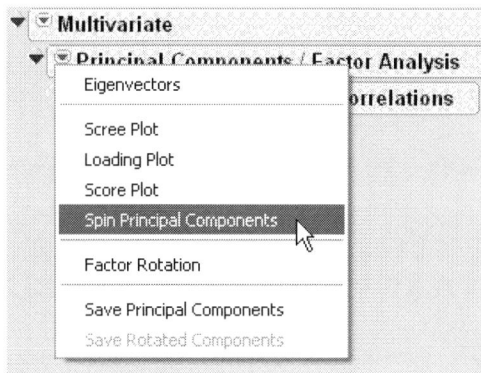

The results are shown in Figure 9.5. Note that because the data are highly correlated, the scatter in the points runs in a narrow ellipse whose principal axis is oriented in the direction marked Prin1.

Figure 9.5 Two Correlated Variables with Principal Components

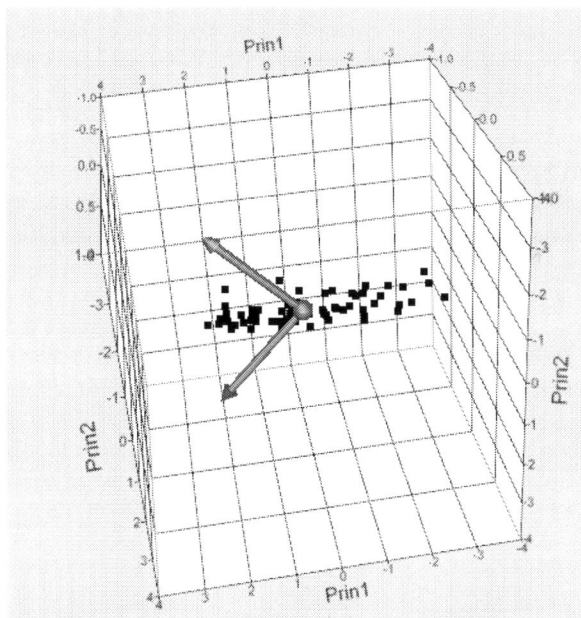

To see the greatest variation of the data in one dimension,

⍵ Rotate the axis so that the first principal component, Prin1, is horizontal.

Principal components capture the most variation possible in the smallest number of dimensions.

Use principal components to explore all six dimensions with the following steps:

⍵ Select **Analyze > Multivariate Methods > Principal Components**.

⍵ Add all six continuous variables to the **Y, Columns** list.

⍵ Click **OK**.

⍟ Click the red triangle icon in the Principal Components/Factor Analysis title bar and select **Eigenvectors**.

The result is the Principal Components table in Figure 9.6. The cumulative percent row (Cum Percent) shows that the first three principal components account for 97.8% of the six-dimensional variation.

Figure 9.6 Principal Components Report

Multivariate Distance

The basic concept of distance in several dimensions relates to the correlation of the variables. For example, in a Multivariate scatterplot cell for Benzene by Chloroform (Figure 9.3), HYDROQUINONE is located away from the point cloud. This compound is not particularly unusual in either the x or y direction alone, but it is a two-dimensional outlier because of its unusual distance from the strong linear relationship between the two variables. The ellipse is a 95% density contour for a bivariate normal distribution with the means, standard deviations, and correlation estimated from the data. The concept of distance that takes into account the multivariate normal density contours is called *Mahalanobis distance*.

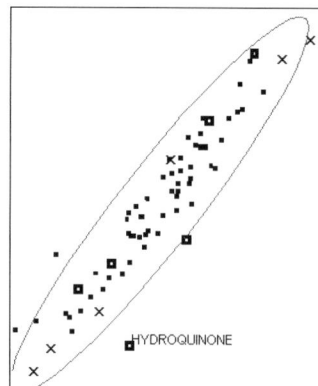

Though only three dimensions can be visualized at a time, the Mahalanobis distance can be calculated for any number of dimensions. To produce a plot of the Mahalanobis distance:

⍟ Select **Outlier Analysis > Mahalanobis Distances** from the menu accessed by the red triangle at the top of the multivariate report.

Figure 9.7 shows the Mahalanobis distance by the row number for each data point. To label these points:

⍟ Select the brush tool ( ) from the tools palette.

⍟ While holding down the Shift key, drag the brush over the points labeled in Figure 9.7. These are the five points with the greatest Mahalanobis distances.

⍟ Select **Rows > Label/Unlabel**.

Figure 9.7 Mahalanobis Distance Plot to See Multivariate Outliers

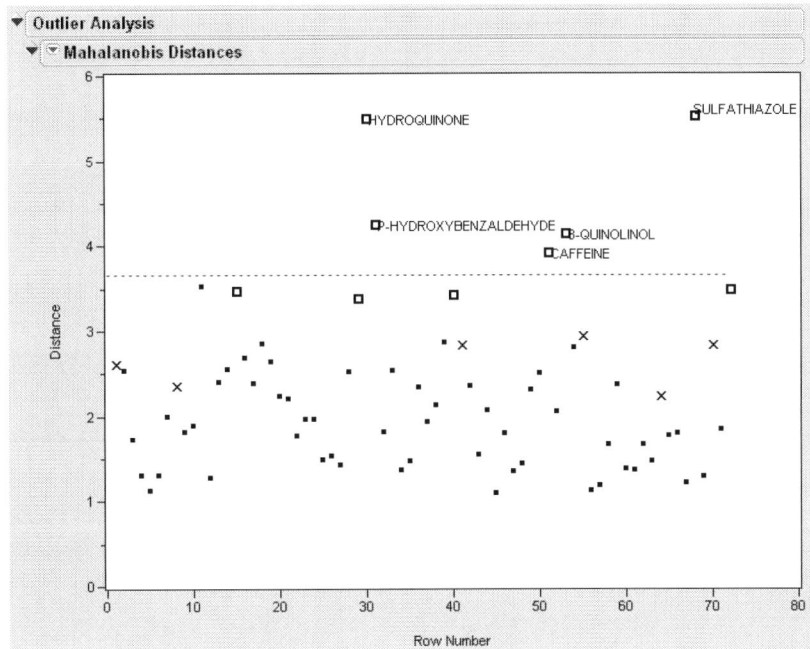

Chapter Summary

In this example, commands from the **Analyze** and **Graph** menus were used for data exploration to locate and identify unusual points. The data were first examined in one dimension using the **Distribution** command and then in two dimensions using the **Multivariate** command to look for unusual points in histograms and scatterplots.

Next, the **Principal Components** command was used to plot three columns at a time. This technique was used to summarize six dimensions and to plot principal component rays. The Principal Components table showed that the first three principal components accounted for more than 97% of the total variation.

Finally, the **Outlier Analysis** command in the Multivariate report produced the Mahalanobis outlier distance plot, which summarizes the points in six dimensions. The multivariate outliers were highlighted and labeled in this multi-dimensional space.

See the chapter "Correlations and Multivariate Techniques," in the *JMP Statistics and Graphics Guide*, for documentation and examples of multivariate analyses. "Three-Dimensional Scatterplots" in the *JMP Statistics and Graphics Guide* documents the 3D plot.

Chapter **10**

Multiple Regression
Examining Multiple Explanations

Multiple regression is the technique of fitting or predicting a response by a linear combination of several regressor variables. The fitting principle is like simple linear regression, but the space of the fit is in three or more dimensions, making it more difficult to visualize. With multiple regressors, there are more opportunities to model the data well, but the process is more complicated.

This chapter begins with an example of a two-regressor fit that includes three dimensional graphics for visualization. The example is then extended to include six regressors (but unfortunately no seven-dimensional graphics to go with it).

Objectives

- Illustrate the concept of a fitting plane using graphical techniques.
- Combine data tables using the **Concatenate** command.
- Explore a three-dimensional version of a leverage plot.

Contents

Aerobic Fitness Data

Aerobic fitness can be evaluated using a special test that measures the oxygen uptake of a person while running on a treadmill for a prescribed distance. However, it would be more economical to evaluate fitness with a formula that predicts oxygen uptake with simpler measurements.

To identify such an equation, runtime and fitness measurements were taken for 31 participants who ran 1.5 miles. The participants' ages were also recorded.

🖰 When you installed JMP, a folder named **Sample Data** was also installed. In that folder is a file named Fitness.jmp. Open Fitness.jmp.

The data are shown in Figure 10.1. For purposes of illustration, certain values of **MaxPulse** and **RunPulse** have been changed from data reported by Rawlings (1988, p. 124).

Figure 10.1 Partial Listing of the Fitness.jmp Data File

		Name	Sex	Age	Weight	Oxy	Runtime	RunPulse	RstPulse	MaxPulse
	1	Donna	F	42	68.15	59.57	8.17	166	40	172
	2	Gracie	F	38	81.87	60.06	8.63	170	48	186
	3	Luanne	F	43	85.84	54.30	8.65	156	45	168
	4	Mimi	F	50	70.87	54.63	8.92	146	48	155
	5	Chris	M	49	81.42	49.16	8.95	180	44	185
	6	Allen	M	38	89.02	49.87	9.22	178	55	180
	7	Nancy	F	49	76.32	48.67	9.40	186	56	188
	8	Patty	F	52	76.32	45.44	9.63	164	48	166
	9	Suzanne	F	57	59.08	50.55	9.93	148	49	155
	10	Teresa	F	51	77.91	46.67	10.00	162	48	168
	11	Bob	M	40	75.07	45.31	10.07	185	62	185
	12	Harriett	F	49	73.37	50.39	10.08	168	67	168
	13	Jane	F	44	73.03	50.54	10.13	168	45	168
	14	Harold	M	48	91.63	46.77	10.25	162	48	164
	15	Sammy	M	54	83.12	51.85	10.33	166	50	170
	16	Buffy	F	52	73.71	45.79	10.47	186	59	188
	17	Trent	M	52	82.78	47.47	10.50	170	53	172
	18	Jackie	F	47	79.15	47.27	10.60	162	47	164

Investigate **Age** and **Runtime** as predictors of oxygen uptake using the Fit Model platform.

To examine a multiple regression model with two effects,

🖰 Highlight **Oxy** in the Select Columns list and click **Y**.

🖰 Highlight both **Age** and **Runtime**.

🖰 Click **Add** to specify them as model effects.

You should now see the completed window shown in Figure 10.2.

Figure 10.2 Completed Fit Model Window for Multiple Regression with Two Effects

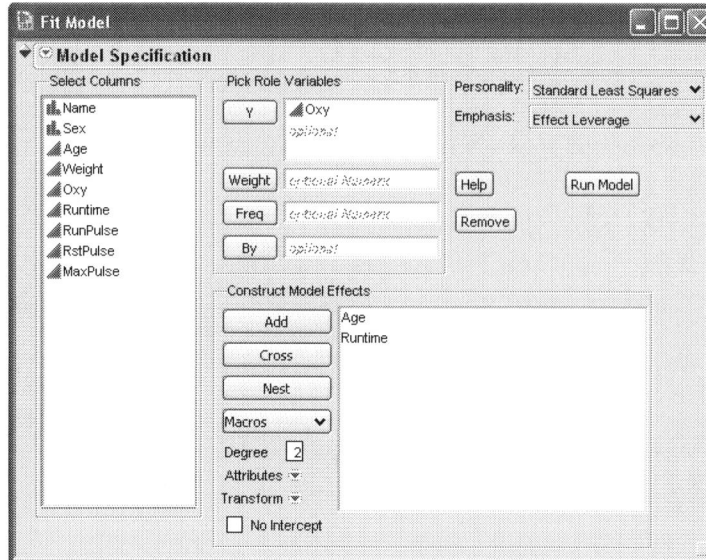

⟜ Click **Run Model**.

You should now see the tables shown in Figure 10.3. These statistical reports are appropriate for a response variable and factor variables that have continuous values.

Figure 10.3 Statistical Text Reports

Clicking the red triangle icon and selecting **Save Columns** displays a list of save commands. To save predicted values and the prediction equation for this model:

⟜ Click the red triangle icon and select **Save Columns > Prediction Formula**.

This command creates a new column in the Fitness data table called Pred Formula Oxy. Its values are the calculated predicted values for the model. To see the column's formula:

⮝ Right-click the Pred Formula Oxy column name and select **Formula**.

The Formula window opens and displays the formula

```
88.4356809 + -0.1509571 * Age + -3.1987736 * Runtime
```

This formula defines a plane of fit for Oxy as a function of Age and Runtime.

⮝ Click **Cancel** to close the window and return to the data table window.

Fitting Plane

JMP can show relationships between Oxy, Runtime, and Age in three dimensions with a surface plot.

⮝ Select **Graph > Surface Plot**.

⮝ Add Oxy and Predicted Formula Oxy as **Columns**, and click **OK**.

⮝ From the **Style** menu for Pred Formula Oxy , select **Needles**. (See Figure 10.4.)

Figure 10.4 Initial View of the Surface Plot of Oxy, Age, and Runtime

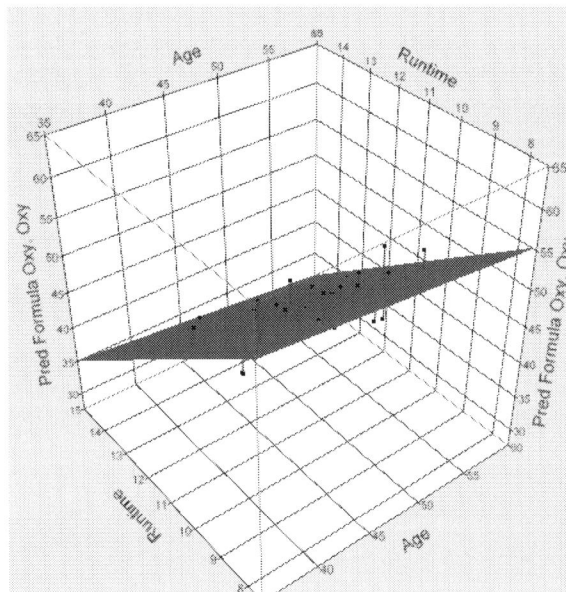

⮝ Click and drag to rotate the plot so it looks like that in Figure 10.5.

Figure 10.5 Observed Points using Age, Oxy, and Runtime with the Predicted Plane of Fit

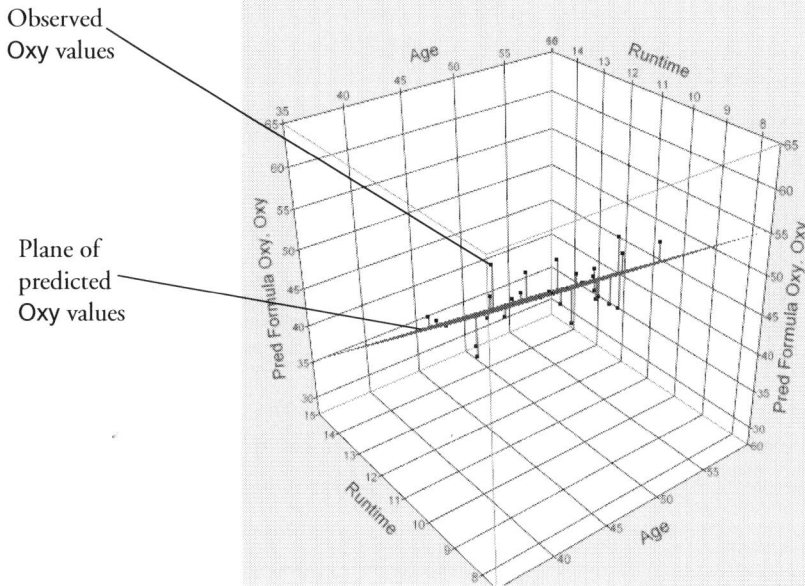

Fit Planes to Test Effects

The example in the previous section showed a plane fit to the whole model. You can also look at hypothesis tests for each regressor to test whether the regressor's parameter is significantly different from zero.

One way to view this test is to evaluate the difference between the current fit and the fit that occurs if the regressor variable is removed from the model.

For example, remove the Runtime variable from the model by following these steps:

↗ First, make sure Fitness.jmp is the active data table.

↗ Select **Analyze > Fit Model**.

↗ In the Fit Model window, select Oxy in the list of variables and click **Y**.

↗ Select Age in the list of variables and click **Add** to add Age as a model effect. Then click **Run Model**.

↗ Click the red triangle icon in the Response Oxy report and select **Save Columns > Prediction Formula**.

The new predicted column (labeled Pred Formula Oxy 2) is calculated using the formula

    62.4229492 + -0.3156031*Age

To compare this fitted line with the plane in the previous example,

↗ Select **Graph > Surface Plot**.

↗ Add Oxy, Pred Formula Oxy, and Pred Formula Oxy 2 as **Columns** and click **OK**.

🖰 In the **Point Response** column drop-down menu, select Oxy for both Pred Formula Oxy and Pred Formula Oxy 2.

🖰 In the **Style** drop-down menu, select **Needles** for both Pred Formula Oxy and Pred Formula Oxy 2.

🖰 In the **Surface** drop-down menu, select **Both Sides** for Pred Formula Oxy and Pred Formula Oxy 2.

Both this grid and the one in Figure 10.5 represent least squares regression planes, but this plane has a slope of zero in the orientation of the Runtime axis. Figure 10.6 shows the plot from an angle.

Figure 10.6 Three-Dimensional Plot with Regression Planes

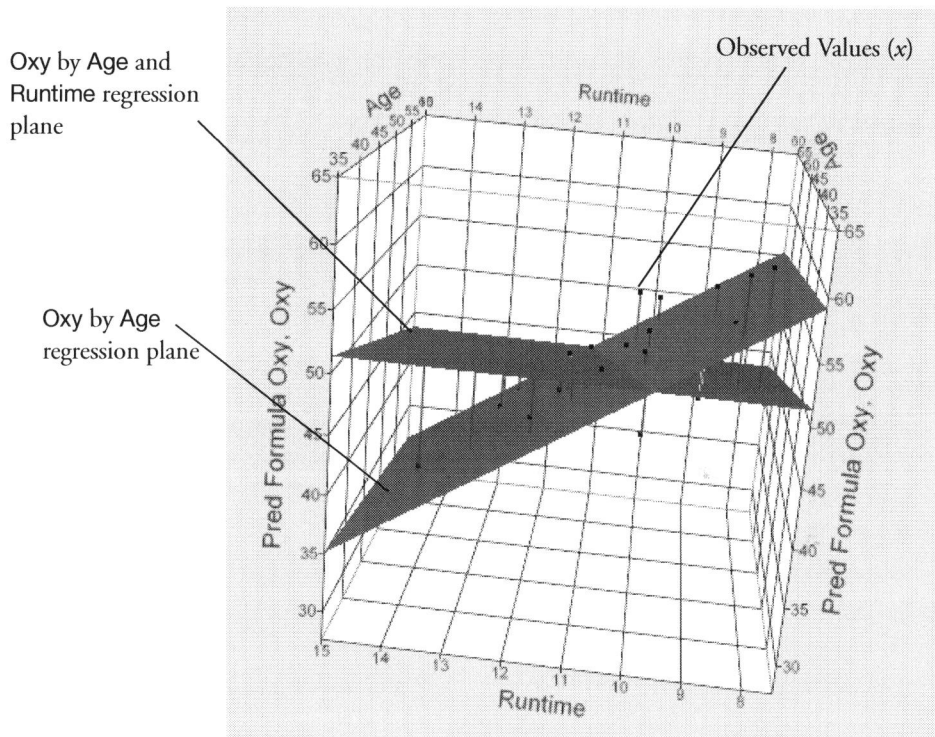

🖰 Click and drag to rotate the plot.

A rotated view is shown in Figure 10.7. Notice this subset (Age-only model) regression showing as a line instead of a plane. The view is *edge-on* for Runtime, which eliminates it from the visual model.

Figure 10.7 Comparison of Three-Dimensional Views

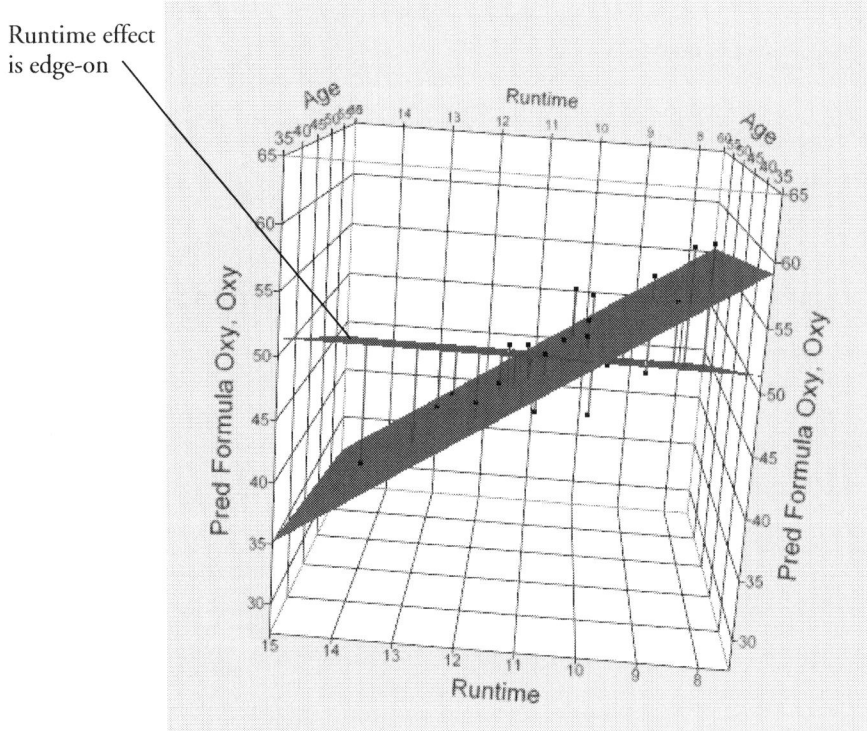

Whole Model Tests

The leverage plot in Figure 10.8 shows joint test of the Age, Weight, Runtime, RunPulse, and MaxPulse effects in the model. This plot compares the full model with the model containing the intercept that fits the overall response mean only. This leverage plot is formed by plotting the actual observed values on the y-axis and the values predicted by the whole model on the x-axis. The residual for the subset model is the distance from a point to the horizontal line drawn at the sample mean.

Figure 10.8 Leverage Plot for the Whole Model (Age and Runtime)

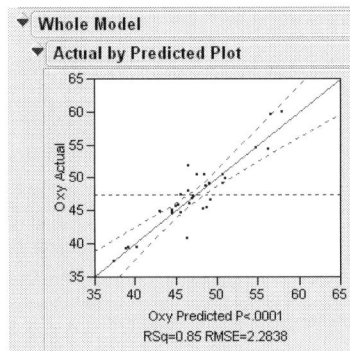

More and More Regressors

It's easy to visualize two regressors predicting a response by using fitting planes. But how can this be done with more regressors when the analysis requires more than three dimensions? In actuality, the fitting, testing, and leverage plot analyses still work for more regressors.

⊕ Select **Analyze > Fit Model**. Fill the window with Oxy as **Y** and add Age, Weight, Runtime, RunPulse, and MaxPulse as model effects.

⊕ Click **Run Model** to run the model.

In this case, the prediction formula is

```
Oxy = 101.3 - 0.2123 Age - 0.0732 Weight - 2.688 Runtime - 0.3703
RunPulse + 0.3055 MaxPulse
```

Look at the significance of each regressor with t-ratios in the Parameter Estimates table or F-ratios in the Effects Tests table. (See Figure 10.9.) Because each effect has only one parameter, the F-ratios are the squares of the t-ratios, and have the same significance probabilities.

The Age variable seems significant, but Weight does not. The Runtime variable seems highly significant. Both RunPulse and MaxPulse also seem significant, but MaxPulse is less significant than RunPulse.

Figure 10.9 Statistical Tables for Multiple Regression

▼ **Whole Model**

▶ **Actual by Predicted Plot**

▼ **Summary of Fit**

RSquare	0.846848
RSquare Adj	0.816217
Root Mean Square Error	2.283779
Mean of Response	47.37581
Observations (or Sum Wgts)	31

▼ **Analysis of Variance**

Source	DF	Sum of Squares	Mean Square	F Ratio
Model	5	720.99043	144.198	27.6472
Error	25	130.39112	5.216	Prob > F
C. Total	30	851.38154		<.0001*

▼ **Parameter Estimates**

Term	Estimate	Std Error	t Ratio	Prob>|t|
Intercept	101.34768	11.86665	8.54	<.0001*
Age	-0.212322	0.094388	-2.25	0.0335*
Weight	-0.073205	0.053611	-1.37	0.1843
Runtime	-2.688436	0.34207	-7.86	<.0001*
RunPulse	-0.370263	0.117723	-3.15	0.0042*
MaxPulse	0.3055336	0.134541	2.27	0.0320*

▼ **Effect Tests**

Source	Nparm	DF	Sum of Squares	F Ratio	Prob > F
Age	1	1	26.39162	5.0601	0.0335*
Weight	1	1	9.72475	1.8645	0.1843
Runtime	1	1	322.16433	61.7688	<.0001*
RunPulse	1	1	51.59471	9.8923	0.0042*
MaxPulse	1	1	26.89778	5.1571	0.0320*

Interpreting Leverage Plots

The leverage plots for this example multiple regression model allow visualization of the contribution of each effect. First, look at the whole-model leverage plot, shown in Figure 10.8, of observed versus predicted values. This plot illustrates the test for the whole set of regressors.

The Analysis of Variance table in Figure 10.9 shows a highly significant F corresponding to this plot. The confidence curves show the strong relationship because they cross the horizontal line.

Now examine the leverage plots for the regressors. Each plot illustrates the residuals as they are and as they would be if that regressor were removed from the model.

The confidence curves in the leverage plot for Age in Figure 10.10 show that Age is borderline significant because the curves barely cross the horizontal line of the mean. Note that the significance of the Age effect is 0.03 in the text reports (Figure 10.9), which is only slightly different from the 0.05 confidence curves drawn by JMP.

The leverage plot for Weight shows that the effect is not significant. The confidence curves do not cross the horizontal line of the mean.

Figure 10.10 Leverage Plots for the Age and Weight Effects

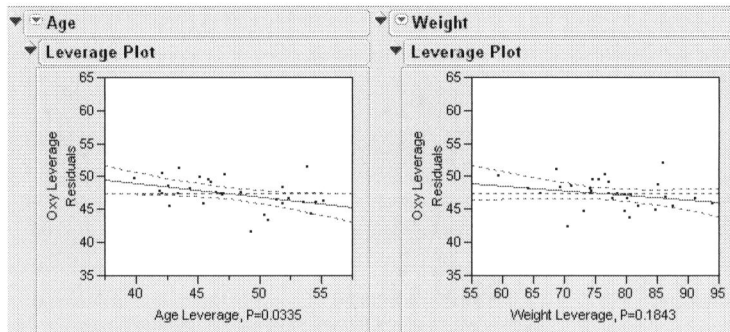

The leverage plot for Runtime shows that Runtime is the most significant of all the regressors. The Runtime leverage line and its confidence curves cross the horizontal mean at a steep angle.

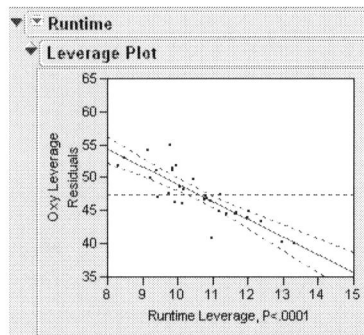

The leverage plots for RunPulse and MaxPulse shown in Figure 10.11 are similar. Each is somewhat shrunken on the x-axis. This indicates that other variables are related in a strong, linear fashion to these two regressors, which means the two effects are strongly correlated with each other.

Figure 10.11 Leverage Plots for the RunPulse and MaxPulse Effects

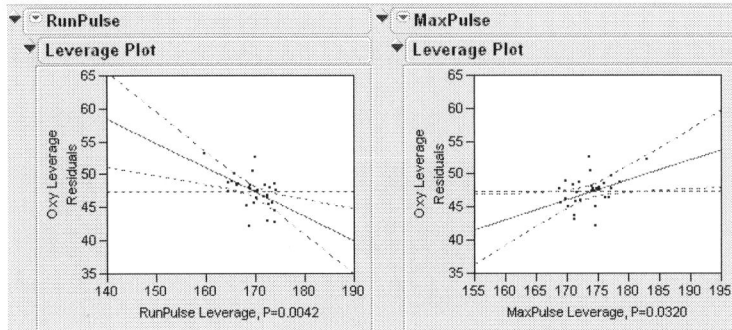

Collinearity

When two or more regressors have a strong correlation, they are said to be *collinear*. These regression points occupy a narrow band showing their linear relationship. When a plane is fit representing collinear regressors, the plane fits the points well in the direction where they are widely scattered. However, in the direction where the scatter is very narrow, the fit is weak and the plane is unstable.

In text reports, this phenomenon translates into high standard errors for the parameter estimates and potentially high values for the parameter estimates themselves. This occurs because a small random error in the narrow direction can have a huge effect on the slope of the corresponding fitting plane. An indication of collinearity in leverage plots is when the points tend to collapse toward the center of the plot in the *x* direction.

The Longley.jmp example shows collinearity geometrically in the strongly correlated regressors, X1 and X2. To examine these regressors, examine Figure 10.12, which shows rotated views of the regression planes. They illustrate a regression of X1 on Y, X2 on Y and both on Y. Most of the points are near the intersection of the three planes. All three planes fit the data well, but their vastly different slopes show that the hold is unstable.

Geometrically, collinearity between two regressors means that the points they represent do not spread out in *x* space enough to provide stable support for a plane. Instead, the points cluster around the center causing the plane to be unstable. The regressors act as substitutes for each other to define one direction redundantly. This is cured by dropping one of the collinear regressors from the model. In this case, drop either X1 or X2 from the model, since both measure essentially the same thing.

Figure 10.12 Comparison of RunPulse and MaxPulse Effects

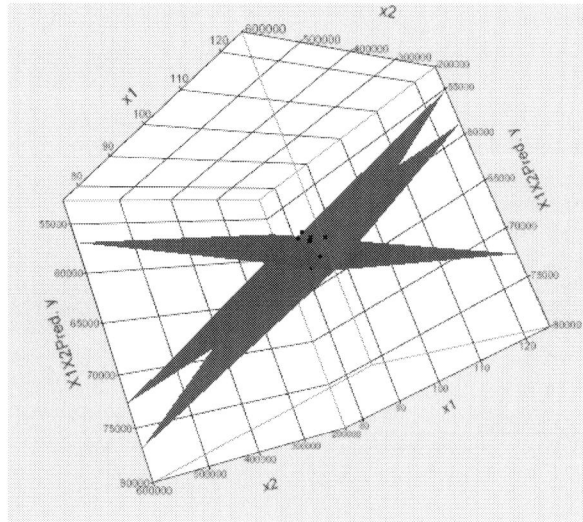

Chapter Summary

Multiple regression uses the same fitting principle as simple regression, but accounting for significance is more subtle. Each regressor opens a new dimension for fitting a hyperplane, and its significance is tested by how much the fit suffers in its absence. When regressors correlate to each other, they are said to be collinear, and they define directions where the fitting hyperplane is not well supported.

References

Becker, R.A., and Cleveland, W.S. (1987), "Brushing Scatterplots," *Technometrics*, 29, 2.

Belsley, D.A., Kuh, E., and Welsch, R.E. (1980), *Regression Diagnostics*, New York: John Wiley & Sons.

Box, G.E.P., Hunter, W.G., and Hunter, J.S. (1978), *Statistics for Experimenters*, New York: John Wiley & Sons, Inc.

Daniel C. and Wood, F. (1980), *Fitting Equations to Data*, Revised Edition, New York: John Wiley & Sons, Inc.

Draper, N. and Smith, H. (1981), *Applied Regression Analysis*, 2nd Edition, New York: John Wiley & Sons, Inc.

Eppright, E.S., Fox, H.M., Fryer, B.A., Lamkin, G.H., Vivian, V.M., and Fuller, E.S. (1972), "Nutrition of Infants and Preschool Children in the North Central Region of the United States of America," *World Review of Nutrition and Dietetics*, 14.

Eubank, R. L., (1988), *Spline Smoothing and Nonparametric Regression*, New York: Marcel Dekker.

Gabriel, K.R. (1982), "Biplot," *Encyclopedia of Statistical Sciences*, Volume 1, Kotz and Johnson editors, New York: John Wiley & Sons, Inc.

Hartigan J.A. and B. Kleiner (1981), "Mosaics for Contingency Tables," *Proceedings of the 13th Symposium on the Interface between Computer Science and Statistics,* W. F. Eddy editor, New York: Springer.

Hawkins, D.M., (1974), "The Detection of Errors in Multivariate Data Using Koehler, Grigorus, Dunn (1988), "The Relationship Between Chemical Structure and the Logarithm of the Partition," *QSAR*, 7.

Koehler, M.G., Grigorus, S., and Dunn, J.D. (1988), "The Relationship Between Chemical Structure and the Logarithm of the Partition Coefficient," Quantitative Structure Activity Relationships, 7.

Leven, J. R., Serlin, R. C., and Webne-Behrman, L. (1989), "Analysis of Variance Through Simple Correlation," *American Statistician*, 43.

Mosteller, F. and Tukey, J.W. (1977), *Data Analysis and Regression,* Reading Mass: Addison-Wesley.

Rawlings, J. O., Pantula, S.G., and Dickey, D.A. (1998), *Applied Regression Analysis: A Research Tool-2nd ed.*, New York, NY: Springer-Verlag New York Inc.

Sall, J. P. (1990), "Leverage Plots for General Linear Hypotheses," *American Statistician*, 308-315.

SAS Institute (1987), *SAS/Stat Guide for Personal Computers, Version 6 Edition*, Cary NC: SAS Institute Inc.

Snedecor, G.W. and Cochran, W.G. (1967), *Statistical Methods*, Ames Iowa: Iowa State University Press.

Winer, B.J. (1971), *Statistical Principals in Experimental Design*, 2nd Edition, New York: McGraw-Hill, Inc.

Index

T

W-Z

Your Turn

We welcome your feedback.

- If you have comments about this book, please send them to yourturn@sas.com. Include the full title and page numbers (if applicable).
- If you have comments about the software, please send them to suggest@sas.com.

SAS® Publishing Delivers!

Whether you are new to the work force or an experienced professional, you need to distinguish yourself in this rapidly changing and competitive job market. SAS® Publishing provides you with a wide range of resources to help you set yourself apart. Visit us online at support.sas.com/bookstore.

SAS® Press

Need to learn the basics? Struggling with a programming problem? You'll find the expert answers that you need in example-rich books from SAS Press. Written by experienced SAS professionals from around the world, SAS Press books deliver real-world insights on a broad range of topics for all skill levels.

support.sas.com/saspress

SAS® Documentation

To successfully implement applications using SAS software, companies in every industry and on every continent all turn to the one source for accurate, timely, and reliable information: SAS documentation. We currently produce the following types of reference documentation to improve your work experience:

- Online help that is built into the software.
- Tutorials that are integrated into the product.
- Reference documentation delivered in HTML and PDF – **free** on the Web.
- Hard-copy books.

support.sas.com/publishing

SAS® Publishing News

Subscribe to SAS Publishing News to receive up-to-date information about all new SAS titles, author podcasts, and new Web site features via e-mail. Complete instructions on how to subscribe, as well as access to past issues, are available at our Web site.

support.sas.com/spn

§.sas® | THE POWER TO KNOW®

7137056R0

Made in the USA
Lexington, KY
24 October 2010